INTERACT
ARCHITEC

ADAPTIVE WORLD

Michael Fox, Editor

Princeton Architectural Press, New York

FOREWORD

Looking through the work in this book, I am immediately struck by how much has changed in the landscape of interactive architecture over the past five years. Where Michael Fox's first book on the field, written with Miles Kemp, presented a vast and promising array of protoarchitectural projects, here we see how quickly the field is maturing, evolving into ambitious permanent and semipermanent installations.

I believe this collection of work reflects an exciting opportunity, in a world of rapid technological change, for architecture to expand its practice and reaffirm itself as the melting pot of the arts and sciences. At the Interactive Architecture Lab, we're finding that sensory and responsive technologies expose new and surprising ways to make connections across disparate fields, such as between robotics and the performing arts, wearable computing and perceptual sciences, biology and the visual arts, and artificial intelligence and digital fabrication.[1] And all of this is happening within the wider context of whirlwind progress in robotics that promises driverless cars, autonomous flying vehicles, and seemingly endless other computerized forms that will soon share our built environment.

As these technologies become part of our design tool kit, our typical aesthetic considerations of space, form, and surface expand to encompass concerns of the aesthetics of behavior. Increasingly active, responsive, and kinetic, the material of the built environment is being animated in the truest sense of the word. Architecture imbued with autonomy, an uncanny sense of life, challenges us to look beyond design disciplines to understand the perceptual, emotional, and social effects of these pervasive technologies.

Puppetry, an ancient art with a rich, albeit poorly recorded, history, offers us a performative perspective. The trained puppeteer, a conjurer of what Roman Paska calls the "theatre of possession,"

exploits the spectator's sensitivity to the subtitles of motion cues.[2] Through careful manipulation of rods and strings with rhythmic motion, the essential breath of the puppet manifests as life: the speed, duration, acceleration, and deceleration profiles of motion conveying inner emotional states. With the subtlest of changes in rhythm, the puppeteer conveys character in matter, with causal and narrative relationships born of the performance of objects and their environment. Even with little more than breath, a thrilling, complex, and challenging set of aesthetic opportunities can be harnessed, and as the range of gestures grows, the potential to imbue objects and, indeed, architecture with life and character seems endless.

Throughout the history of natural and social sciences, from Aristotle's theories of motion as the exclusive characteristic of living things and the early anthropological studies of primitive cultures to the foundations of perceptual and cognitive psychology, we find the association of movement with life itself to be deep and universal. Today's advancements in medical imaging are revealing the neurological roots of this association; it's in the very architecture of the social human brain. At its core is the instinct to anthropomorphize the nonhuman, whether animal, inanimate object, or even natural phenomenon—to project personality onto other entities as a means of better relating to them.

So as the worlds of architecture and robotics collide, offering new motive and spatial forms of interaction, the cerebral processes of human social relationships are irresistibly stimulated. This renders in strange and uncanny terms a built environment that may viscerally feel worthy of our care and consideration in ways that inanimate matter cannot. We may come to perceive an anthropomorphized architecture as responsible for its own behavior and perhaps even deserving of punishment or reward. As the eminent robotics engineer and philosopher

Rodney Brooks has commented, "I'll eventually feel we have succeeded if we ever get to the point where people feel bad about switching Cog X off."[3] The surreal psychological and social effects are impossible to fully anticipate. What seems certain is that we are at present ill-equipped either conceptually or technically to understand and craft this new aesthetic of behavior.

But then this is really what is so fascinating and what compels us to pursue the potential of interactive architecture. We can't forget that machines are not an entirely new preoccupation. Vitruvius dedicated an entire book of his treaties to machines. Greek mythology told us of Daedalus, the architect of the labyrinth at Knossos, who also crafted magnificent mechanical statues. And by the Renaissance, automatons flourished to act as centerpieces in royal courts and town squares alike. The Florentine Francini brothers' hydraulic statues of Saint-Germain-en-Laye famously inspired Descartes to construct his own automaton; a pursuit of understanding that challenged the relationship between the body-machine and the mind-soul, it animated not only cogs and levers, but also the very foundations of Western philosophy.

There is something essentially human about making things that come to life, whether they are mechanical, robotic, cyborg, or architectural. It touches upon a human fascination with looking for life in the inanimate on the one hand and a yearning to play god on the other.[4] The words magic and machine share the common etymological root *magh*, meaning "to be able" and "to have power." We have a compulsion to understand other living things and to imitate them. Art itself may be understood as product of the human need to remake the world in search of deeper understanding.

Let's continue that search through code, electronics, networks, mechanics, materials, and novel methods of fabrication. As distinctions between designer and engineer, fabricator and philosopher dissolve, architecture occupies the foreground as a space of radical multidisciplinary or even antidisciplinary practice. Let this be an opportunity for architects to ask questions not only about the future of our homes, workplaces, and public spaces, but also about what it is to be human, our social nature, the future of communities mediated by technology, and our changing relationship to the inanimate and animate world around us.

Ruairi Glynn, 2016

INTRODUCTION: CATALYST DESIGN IN A CONNECTED WORLD

As we embrace a world in which the lines between the physical and the digital are increasingly blurred, we see a maturing vision for architecture that actively participates in our lives. In the few years since the original *Interactive Architecture* was published, a number of projects have been built at scales that both move beyond the scope of the architectural exhibit as test bed and push the boundaries of our thinking in terms of material performance, connectivity, and control. Our architectural surroundings have become so inextricably tied to technological trends that the two ultimately and simultaneously respond to and define each other. The promise of ubiquitous computing has secured a permanent foothold in our lives and has begun to infiltrate our devices and objects as well as our buildings and environments. Such is our physical world: not just digital but also seamlessly networked and connected, an architectural world that is a direct participant in our lives. Bill Gates once predicted that by the end of the first decade of the twenty-first century there would be nothing untouched by the digital.[1] By the end of the second decade, states the interaction designer Behnaz Farahi Bouzanjani, this impact will arguably have become so pervasive that computation will not be noticeable anymore.[2] The subject of this book is how architectural design integrates and negotiates the digital; in our contemporary context, this is nothing short of reciprocal innovation. This book surveys the rapidly evolving landscape of projects and trends that are finally catching up with the past. As a matter of definition, interactive architectural environments are built upon the convergence of embedded computation and a physical counterpart that satisfies adaptation within the framework of interaction. It encompasses both buildings and environments that have been designed to respond, adapt, change, and come to life.

Young designers have started to realize that it is possible to build anything they can imagine.

Sensors available today can discern almost anything, from complex gestures to CO_2 emissions to hair color. An interconnected digital world means, in addition to having sensory perception, that data sets—ranging from Internet usage to traffic patterns and crowd behaviors—can be drivers of interactive buildings or environments. Courses in robotic prototyping and interaction are commonly taught in today's architecture programs, with contextual subjects ranging from urban social issues to practical sustainability. Perhaps equally as important as the rapid advance of such technologies is the fact that both robotics and interaction are technically and economically accessible. The requisite technologies are simple enough to enable designers who are not experts in computer science to prototype their ideas in an affordable way and communicate their design intent. Architects and designers are not expected, as on exhibit-scale projects, to execute their interactive designs alone; they are expected, rather, to possess enough foundational knowledge in the area to contribute. In the same way, while architects need to learn structural engineering in school and, until recently, have been required to pass a special section on structures for the professional licensing exam, it is rarely assumed that architects will do the structural calculations for the buildings they design; that work is carried out by professional structural engineers.

The field is fresh with original ideas, illuminated by the built prototypes and architectural projects illustrated in this book. Driven by the applications, these genuinely new developments and ideas will rapidly foster advanced thinking within the discipline; yet it is important to understand that their foundations have been around for quite some time, dating back nearly thirty years.

CATCHING UP WITH THE PAST
Essentially, the theoretical work of a number of people working in cybernetics in the early 1960s laid

most of the groundwork for the projects highlighted in this book. During this time, Gordon Pask, Norbert Wiener, and other cyberneticians made advancements toward understanding and identifying the field of interactive architecture by formulating their theories on the topic. Pask's conversation theory informed much of the original development in interactive architecture, basically establishing a model by which architects interpreted spaces and users as complete feedback systems.[3] Cybernetic theory continued to be developed into the late sixties and early seventies by the likes of Warren Brody, Nicholas Negroponte, Charles Eastman, Andrew Rabeneck, and others, who expanded upon the earlier ideas of Pask and Wiener.

These early philosophies were then picked up by a few architects who solidly translated them into the arena of architecture. This work generally remained in the realm of paper architecture, however. Cedric Price was perhaps the most influential of the early architects to adopt the initial theoretical work in cybernetics, expanding it into the architectural concept of anticipatory architecture. John Frazer extended Price's ideas in positing that architecture should be a "living,

evolving thing."[4] Yet it is important to understand that while architects were developing these concepts, areas of digital computation and human interaction were advancing in parallel fashion within the sphere of computer science. From this work, fields such as intelligent environments (IE) were formed to study spaces with embedded computation and communication technologies, in turn creating spaces that bring computation into the physical world. Intelligent environments are defined as spaces in which computation is seamlessly used to enhance ordinary activity.[5] Numerous technologies were developed in this area to deal with sensory perception and human behaviors, but the corresponding architecture was always secondary as it was developed under the mantra of "seamlessly embedded computation." In other words, there was very little architectural involvement in the developing field of computationally enhanced environments. Corporate interests, meanwhile, established market-driven interests that played a major role in computationally enhanced environments through the development of numerous market-driven products and systems that directly involved users in the real world. In the 1990s there were "smart home" and "smart workplace" projects being initiated at every turn that relished the newly available technologies. For the first time, wireless networks, embedded computation, and sensor effectors became both technologically and economically feasible to implement by computer science. This feasibility fueled experimentation with many of the ideas of the previously mentioned visionary architects and theoreticians, who had been stifled by the technological and economic hurdles of their day. We are now at a time when the economics of affordable computational hardware and increased aptitude for integrating computational intelligence into our environments has become accessible to architects.

A CONNECTED WORLD

The influence of technological and economic feasibility within a connected world has resulted in the explosion of current exploration with the foundations of interaction design in architecture. The Internet of Things (IoT) has quite rapidly come to define the technological context of interactive design as all-inclusive, existing within this connectedness in a way that affects essentially everything, from graphics to objects to buildings to cities. To use an architectural analogy, the theoretical foundations have a structure that resides in the connected worlds of Web and mobile and spatial interfacing, and they are still evolving. Theories of a connected architectural world existed long before mobile devices and Web-interface technologies changed every aspect of our lives and created the discipline of interaction design.

While the first wave of connectivity focused on human-to-human communication, the current focus is on connected things and devices, which extends naturally to buildings, cities, and global environments. There are approximately one billion websites and about five billion mobile phones, while there are approximately fifty billion smart devices.[6] It is the goal (and responsibility) of the Internet of Things to connect them in a meaningful way.[7] These intelligent things are everywhere in our lives, and many of them are already seamlessly embedded in our architecture, from our kitchen appliances and our HVAC (heating, ventilation, and air conditioning) systems to our home entertainment systems. For the time being, most of them are weakly connected at best. Today the Internet supports hundreds of protocols, and it will support hundreds more. While the world struggles with a protocol platform, the battle over which protocol will prevail is being waged at a staggering commercial cost, often referred to as the "protocol wars."

There are numerous contenders in the game—the IoT needs many. Currently heading the pack are CoAP, MQTT, and XMPP. The important difference between them lies in the distinction of application or the class of use. Devices must communicate with each other (D2D); device data must then be collected and sent to the server infrastructure (D2S). That server infrastructure has to share device data (S2S), possibly providing it back to devices, to analysis programs, or to people.[8] Eventually, all of these connected things will need an infrastructure to enable them to work together. There are a number of companies currently vying for position; their approaches range from cloud-based software (with precedent in things like vending-machine inventory and engine maintenance) to ultra-narrowband radio transmissions. More than likely, the familiar tech trend will prevail: all of the novel small companies with their individual takes on a similar problem will be pounced on by Apple, Microsoft, or Google, who will then take the best of each of them and create their own platforms. The goal of these big companies is to lock everything into their powerful existing systems.

There is currently a need for standardization to avoid having one of the big companies determine this eventual fate, which could indeed result in a nightmare where nothing works outside a proprietary system. "By embracing open standards, we can ensure we won't be locked out of a device or forced to use only one type of connector at the whim of a single company," says Mat Honan in *WIRED* magazine.[9] We have in the past embraced such standards, whereby almost all mobile devices already communicate via the same Bluetooth wireless standard. The point is that every existing company needs to rally behind a common standard—and do it soon. Scott Fisher, the founding chair of the Interdivisional Media Arts + Practice (iMAP) PhD program of the School of Cinematic Arts at the University of Southern California (USC), observes: "The growing number of ubiquitous and embedded computing technologies introduces a new paradigm for how we interact with the built environment, while mobile and pervasive devices offer new possibilities for sensing and communicating with buildings and objects in the physical world. These technologies are used not only for collecting and providing data, but also as a way to animate and collectively augment the world around us."[10] Interactions are no longer limited to those of people interacting with an object, environment, or building, but can now be carried out as part of a larger ecosystem of connected objects, environments, and buildings that autonomously interact with each other. Much of the work at iMAP has been focused on creating interactive architectural environments in which the buildings themselves become storytelling characters. As the design researcher Jen Stein states, "By inviting inhabitants to engage with both the building and other inhabitants, we have introduced a new paradigm for place making within an animated, interactive environment."[11]

Usman Haque is a designer with a background in interactive architecture who has led the way in developing a scalable platform for connectivity with Pachube, which provides a platform for connecting various sensor data and visualizations. Through the development of an Extended Environmental Markup Language (EEML), the platform handles both Web-based and mobile applications for the sharing of sensory and environmental data in real time. Pachube was acquired by COSM, then acquired by Xively (LogMeIn), which encourages open digital ecosystems, connecting more than 250 million devices, including electricity meters, weather stations, building-management systems, air-quality stations, and biosensors, to name just a few.

Architectural applications are iterative in such a connected context. The sensors and robotic components are now both affordable and simple enough for the design community to access; and all of the parts can easily be digitally connected to each other. Designing interactive architecture in particular is not inventing so much as understanding what technology exists and extrapolating from it to suit an architectural vision. In this respect, the designers of buildings, cities, and larger interconnected ecosystems have learned a great deal from the rapidly developing field of tangible interaction, essentially an alternate vision for interfacing that was developed to bring computing back into the real world. Tangible user interfaces were envisioned as an alternative to graphical displays—an alternative that would bring some of the richness of the interaction we have with physical devices back into our interaction with digital content.[12] In contrast, the field of industrial design came to engage with tangible interaction out of necessity as appliances became progressively "intelligent," containing more and more electronic and digital components.[13] Broadly, tangible interaction encompasses user interfaces and interaction approaches that emphasize the sensory appeal and materiality of the interface, the physical embodiment of data, whole-body interaction, and the embedding of the interface and the users' interaction in real spaces and contexts.

Tangible interaction is a highly interdisciplinary area. It spans a variety of perspectives, among them human-computer interaction (HCI) and interaction design, but specializes in interfaces or systems that are in some way physically embodied. Furthermore, it has connections with product and industrial design, arts, and architecture. In a sense, interactive architecture falls under the umbrella of tangible interaction along with environments and physical-artifact, product, and industrial design, only the scale is often much larger.

Although tangible interaction typically deals with the interfacing of objects and artifacts, the connected capabilities have opened up a wealth of possibilities not only at the scale of the building, but also in the city and beyond. One of the pioneers in this area has been the MIT SENSEable City lab, led by Carlo Ratti (who also comes from a background in architecture). The lab has done extensive research into how real-time data generated by sensors, mobile phones, and other ubiquitous technologies can teach us how cities are used and how new technologies will ultimately redefine the urban landscape. Ratti argues that urban planning is not just about cities, but about understanding the combination of physical and digital. Ratti says "[T]he interesting thing is that now the machine, the computer, is becoming the city. The city has become the interface—to retrieve information, to meet other people, to do all the things happening now with this mixing of bits and atoms. So it's this new exciting equation, putting together people, space, and technology."[14]

Additionally, the Situated Technologies initiative, led by Omar Kahn, Trebor Scholz, and Mark Shepard, has had a major influence in this area through symposia, competitions, and publications. The initiative, which emerges from architecture as opposed to computer science, takes into account the social dimension of ubiquitous computing.

It is impossible to predict how quickly interactive architecture will be widely executed or what standards and protocols will work their way to the fore. Yet the projects in this book illustrate that such standards and protocols are becoming an inevitable and completely integral part of how we will make our buildings environments and cities in the future. The platform is ripe to foster unique applications tied to our living trends, which both affect and are affected by digital technology. The chapters that follow document a select number of pioneering projects that are defining the future of interaction. The projects, which are illuminated firsthand by images and text from the architects and designers who brought them to life, give insight into the technology and construction that will be an inevitable and integral part of how we think about architecture. Within a profession recently dominated by a discourse of style, we have begun to detect a shift away from questions of representation and images toward processes and behaviors.[15] Specific categorical areas have consequently come to the fore as designers have forged ahead to pioneer this new area of design. Therefore the projects are organized not by how they are made or how they look, but rather according to what they do: exhilarate, communicate, mediate, evolve, and catalyze.

Michael Fox, 2016

EXHILARA

Definition: to make cheerful and excited; enliven, elate, move
Related words: arouse, incite, inspire, provoke, stimulate; bewitch, captivate, charm, delight, enchant, enthrall, hypnotize, mesmerize, rivet, spellbind; interest, intrigue, tantalize

Merriam-Webster OnLine, s.v. "exhilarate," accessed March 25, 2015,
http://www.merriam-webster.com/dictionary/exhilarate.

TE

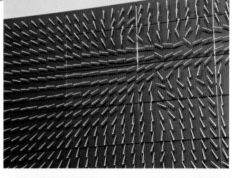

If an environment could adapt to our desires, it would have the ability to shape our experience. The projects in this chapter highlight the emotive possibilities of interactive architecture. There is a great deal of built precedence in interactive applications geared toward the evocation of feeling, ranging from those that simply provide pleasure to those that enable social engagement and contribute educational benefits. In the public realm, artistic structures such as sculptures, fountains, and facades have adopted interactivity as a vital component, inherent to the works in order to capture an audience. Museums as well have rapidly embraced interactivity with respect to the demands of presenting and viewing exhibits and artifacts. Interactivity combined with spatial adaptability can serve well the temporal nature of changing displays and visitors' interaction with them.

Many applications incorporate an educational component whereby kinesthetic learning is combined with entertainment experiences. Such applications enable users to utilize their bodies as well as their minds in collaborative ways. Children seem happy to learn when an entertaining interactive component is involved; being able to control the narrative engages them. While interactive entertainment is rapidly moving into the physical realm, it is a concept born of electronic media. The philosopher Marshall McLuhan lists "three key pleasures" that are uniquely intensified in electronic media: immersion, rapture, and agency. Immersion, he says, is "the sense of being transported to another reality"; rapture is the "entranced attachment to the objects in that reality"; and agency is "the player's delight in having an effect on the electronic world." In the world of entertainment, an engaging environment is by definition successful. Looking at the projects that follow, we see four very different installations that all work successfully to exhilarate.

All of the projects express a critical dimension of time and transformation brought about by physical change. William Zuk and Roger Clark state in their groundbreaking book *Kinetic Architecture* that "our present task is to unfreeze architecture, to make it a fluid, vibrating, changeable backdrop for the varied and constantly changing modes of life. An expanding, contracting, pulsating, changing architecture would reflect life as it is today and therefore be part of it."[1] Kostas Terzidis explains that "deformation, juxtaposition, superimposition, absence, disturbance, and repetition are just a few of the techniques used by architects to express virtual motion and change."[2] He clarifies the polarity that while the form and structure of the average building suggests stability, steadiness, sturdiness, and immobility, the introduction of motion may suggest agility, unpredictability, or uncertainty and may also imply change, anticipation, and liveliness. The integration of motion into the built environment, and the impact of such results upon the aesthetics, design, and performance of buildings, may be of great importance to the field of architecture: "While the aesthetic value of virtual motion may always be a source of inspiration, its physical implementation in buildings and structures may challenge the very nature of what architecture really is."[3] It is important to understand that adaptation in this context is not quite as simple as satisfying needs. The architect Cheng-An Pan states: "Needs and desires change, permitting new options to be employed, allowing greater freedom of geographical movement, accepting personal whim, recognizing changing roles and functions, encouraging personal identity, reflecting mutations in economic levels, and adapting to any change which affects architectural form."[4]

The implications of kinetic architecture touch upon building performance on one hand and aesthetic phenomenology on the other. At an architectural scale, projects often must do both. In the project titled *May/September*, installed by Urbana on the Eskenazi Hospital parking structure facade, the primary goal was to create an exhilarating effect at an urban scale. And yet, as principal Rob Ley points out, the artwork also serves very pragmatically as a visual screen for the ordinary parking structure behind, masking the everyday things one might see there, such as cars, concrete beams, columns, and guardrails. It was required that the piece allow for substantial ventilation, which, as a necessity, worked naturally with the concept. The data from the noise influences the intense visual screen both conceptually and as a functional driver in the image creation. States Ley: "While noise is often understood as an unfortunate by-product of image or sound reproduction, in this case it becomes a modifier of a condition. In the same way that grain can impart a tonal contribution to a photographic image, it can also be synthesized into a numeric data set in such a way that obscurity is controlled." As with many projects in this area, we see the controlled translation of urban phenomenon. In *May/September*, the data, or noise, is technologically sensed and translated to another visual sense so that we can understand it through different patterns.

Translation is also a central theme in the two projects in this section created by Ned Kahn and Charles Sowers. Their work hinges on illuminating unnoticed or invisible phenomena, where the drivers are not only illuminating and, indeed, exhilarating, but are all the more powerful because they teach us about something we might not have been aware of or could not in fact perceive through our senses alone. In some cases this practice involves scaling up the phenomena; in others, it simply means adding a field of passive agents that can be manipulated by forces, making it possible for us to understand. Many (if not most) of the projects in this book rely on a data set of some kind that is sensed and then translated back to the participants. Often the intention is to visualize existing complex patterns and reinterpret them in a medium that is simple enough to comprehend.

Although designers understand and have demonstrated that it is possible to perceive and create data from anything, the power of Kahn's and Sowers's work lies in the exhilarating effect of translating natural phenomena. Their installations don't just make us aware of that which we haven't perceived; they do so in a way that moves us emotionally. For the Technorama Building at the Swiss Science Center, Kahn designed a facade composed of thousands of aluminum panels that move in the air currents to reveal the complex patterns of turbulence in the wind. He is admittedly less concerned with creating a reality than with unveiling the world in perceptible ways. With *Windswept*, Sowers takes a similar approach to translation yet in a particularly low-tech manner, reinterpreting the effects of the wind in a way that rewards extended observation. *Windswept* serves as a scientific instrument of sorts, acting as a discrete window onto a very large phenomenon that until now has been invisible.

The last project in this section takes a high-tech approach that relies on a sympathetic understanding of human behaviors. The same emotive quality brought about in Kahn's and Sowers's portfolios via the translation of natural phenomena is achieved by technological means in the *Reef* project by Rob Ley of Rob Ley Studio and Joshua Stein of Radical Craft. The behavior of the "reef," which responds to people in its space, emulates that of plants and lower-level organisms that are considered responsive but not conscious. Their concern with a nonmechanized, efficient, and fluid movement derives from our emotive interpretation and response to such behaviors. Ley remarks that "*Reef*'s unique exploration of technology shifts from the biomimetic to the biokinetic while liberating and extending architecture's capacity to produce a sense of willfulness." He concludes that "behavior may ultimately be more important than intelligence as we strive for a viable model of interactivity of space and the user."

MAY/SEPTEMBER
Urbana

This project sought to explore parallels between techniques of two-dimensional image construction and the tectonic considerations of building enclosure. Through rigorous examination of digital image manipulation and reproduction techniques, a strategy for the articulation of complex arrangements of patterns and edges across a building facade was developed. The primary conceptual intention of the project was twofold: first, to interrogate the notion of optimization with respect to our contemporary understanding of fabrication, using image as a conceptual link between the efficiency of a digital system and the performance of a real-world tectonic system;

second, to exploit the effects of such a strategy, so that the reduction in resolution of the system enhances the spatial qualities. In this way, optimization becomes an asset to spatial conditioning, rather than a necessary compromise.

The use of digital image mapping and pixel data–extraction techniques in the development of architectural form is neither new nor groundbreaking. In fact, examination of the pixel might be the most rudimentary means of extracting data for digital abstraction and manipulation. For instance, the extraction of color and brightness values from an image as a means of manipulating some systemic design parameter across a field

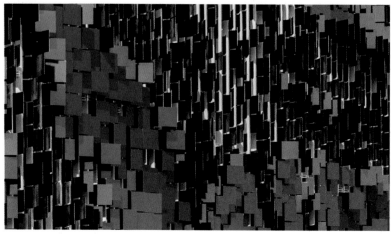

condition features heavily in even the most conceptually superficial attempts at digital design. Yet very rarely is the construction of the digital image itself examined with any degree of rigor. In the development of *May/September*, examination of image began further upstream, at the point at which the image itself is created, taking into account the distortion, abstraction, and optical processing that occurs so that the image in question may be efficiently projected onto a screen. This process of altering the "actual" image data not only increases the efficiency of the digital file but also brings the image closer to its real-world likeness.

Noise, as a concept and as a function, becomes an important tool in the development of the facade. While noise is often understood as an unfortunate by-product of image or sound reproduction, in this case it becomes a modifier of a condition. In the same way that grain can impart a tonal contribution to a photographic image, it can be synthesized into a numeric data set so that obscurity is controlled. The introduction of noise is therefore extremely useful, both in a quantitative sense (by minimizing necessary resources or processing power) and in a qualitative sense (by improving the likeness of the resultant digital image).

Contemporary modes of digital production often strive to capitalize on the potential for unrestrained differentiation in design and output. Mass-customization of components is rapidly becoming the status quo, yet how much of this differentiation is actually necessary for the articulation of gradient spatial conditions? Within the academic discourse of architecture, there is an ongoing trend to embrace infinite variability of components, even though the theoretical ideal of mass-customization remains at odds with the reality of mainstream contemporary construction and fabrication techniques, particularly on an architectural scale. In a typical fabrication context, every variation in assembly component is coupled with a substantial increase in time, labor, and cost—and it would be naive to believe otherwise. This brings the notion of waste to the forefront, both conceptually and pragmatically. As techniques and technology continue to foster the variability of components an inevitable tipping point may be seen on the horizon, at which we start to observe that just because we can doesn't mean we should.

The obvious goal of optimization is to reduce the number of components necessary to create a spatial effect before the system breaks or the overarching design intention is no longer coherent. Beyond this goal, however, there is potential for the process of reductive optimization to improve the spatial integrity of the system. It then becomes both necessary and beneficial to exploit moments of commonality and repetition within the system.

The production of images has naturally followed the historic limitations of print (and, later, digital) technology. Most are familiar with the movable type of the Gutenberg press and the woodblock printing that preceded it. These early technologies—and the improved versions centuries later—allowed for the relatively quick production of printed type. As technology advanced, so did the capability of including images along with type through various mechanically reproducible means. Of particular interest in the project is how binary conditions in the printmaking world resulted in an increased focus on obtaining the most from the least: the ability to produce complex images, including perceived shades of gray or various color hues, while still being produced with a single color. Halftoning is the most commonly used method today and a

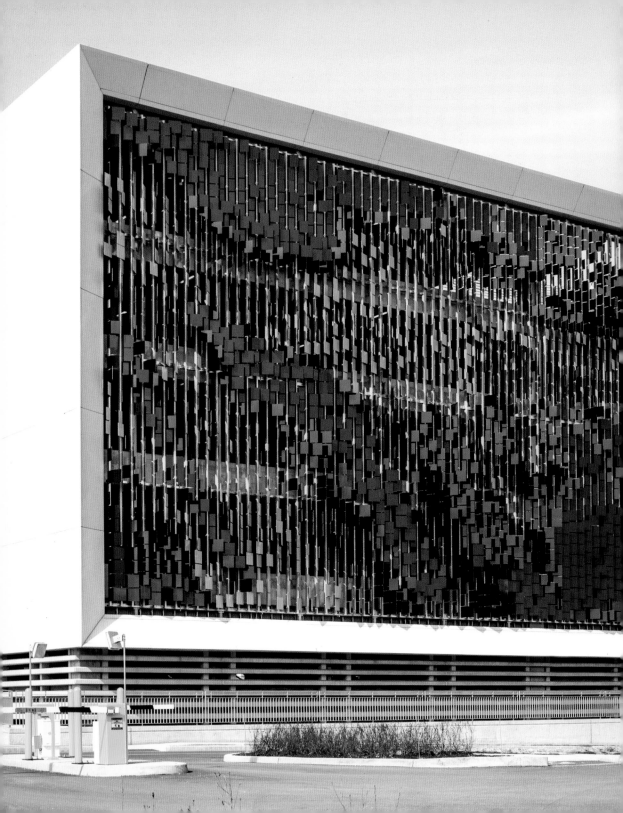

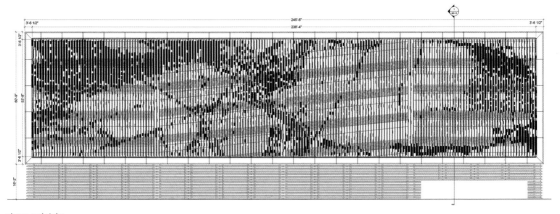

above and right:
Facade elevation
and section detail

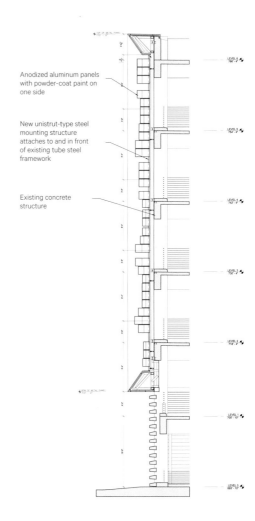

Anodized aluminum panels
with powder-coat paint on
one side

New unistrut-type steel
mounting structure
attaches to and in front
of existing tube steel
framework

Existing concrete
structure

good example of a technique that uses a limited
palette along with spacing and dot sizing to
create a complex image.

May/September derives a noise-based tex-
tural building enclosure by leveraging a palette
of architectural components from the phenom-
enal qualities of dithering and error-diffusion in
printmaking. In this way, three basic typologies
with three sized subsets, along with part mirror-
ing, produce a palette of eighteen unique parts.
These components, along with a yellow/blue
binary color palette, produce a complex, though
nuanced, condition from a relatively small set of
variables. The manipulation and translation of a
limited set of components is a more interesting
challenge than just adding more unique parts.

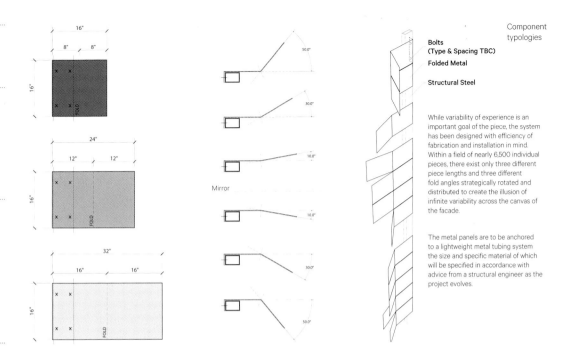

Component typologies

Bolts
(Type & Spacing TBC)

Folded Metal

Structural Steel

While variability of experience is an important goal of the piece, the system has been designed with efficiency of fabrication and installation in mind. Within a field of nearly 6,500 individual pieces, there exist only three different piece lengths and three different fold angles strategically rotated and distributed to create the illusion of infinite variability across the canvas of the facade.

The metal panels are to be anchored to a lightweight metal tubing system the size and specific material of which will be specified in accordance with advice from a structural engineer as the project evolves.

Mirror

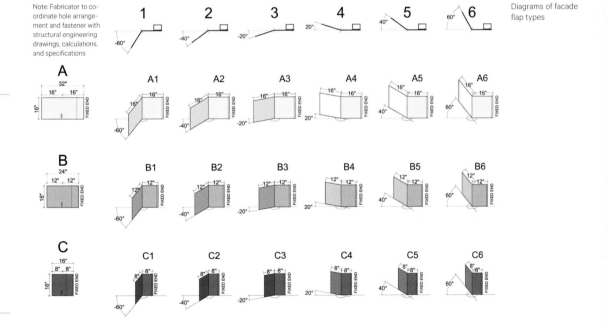

Note: Fabricator to co-ordinate hole arrangement and fastener with structural engineering drawings, calculations, and specifications

Diagrams of facade flap types

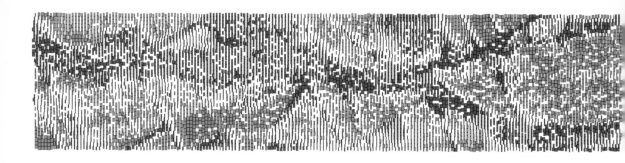

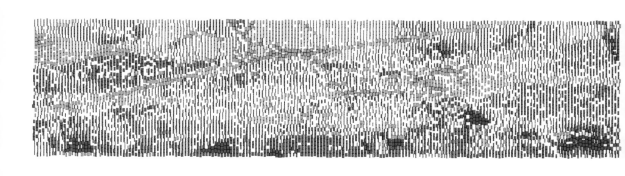

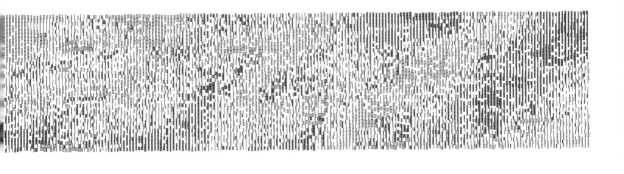

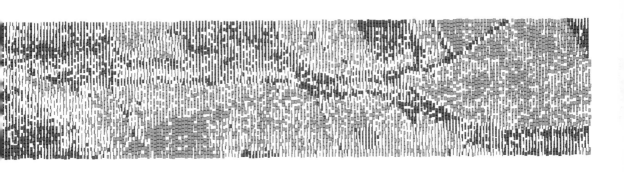

Dithering and noise
studies

TECHNORAMA FACADE
Ned Kahn

In 2002 Ned Kahn worked with the staff of Technorama, the Swiss Science Center, and the institution's architectural office of Durig and Rami to create a six-story facade for the building. The facade comprises thousands of aluminum panels, set in motion by air currents to reveal the complex patterns of turbulence in the wind. For the installation, the entire 220-foot-long (67.1 m) facade of the museum was covered with eighty thousand wind-animated panels. The brushed-aluminum surface of the panels reflects light and color from the sky and the surrounding buildings. As a whole, the facade

is the focal point of the large urban plaza in front of the museum. Each moving element is a three-inch (7.6 cm) square of thin aluminum with a low-friction plastic bearing pressed into the top edge. These bearings ride on stainless-steel axles, held by an aluminum framework to the structural beams of the building. Each element responds uniquely to the wind's forces, but the entire facade displays complex, coordinated movements that coalesce into a rendering of the larger-scale patterns and textures of its passing. The artwork has survived extreme windstorms, ice storms, and more than a decade of constant

Details of facade in
different states

Facade at rest

exposure to the sun without damage or degradation. Applying the same strategy by which trees
deflect damage in high winds with their leaves,
the small surface area of each individual moving
element in the Technorama Facade reduces the
effect of the forces that build up during a storm.

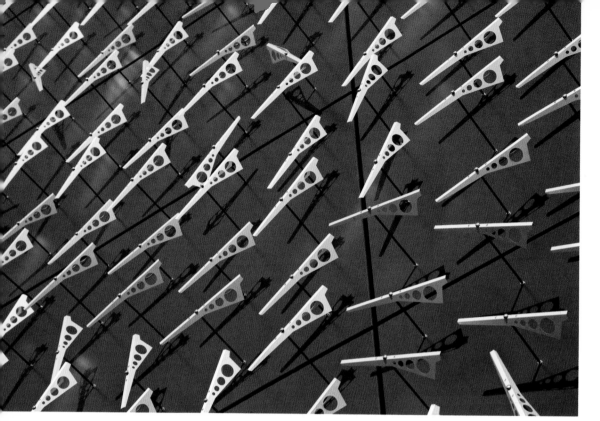

WINDSWEPT
Charles Sowers

Windswept, commissioned by the San Francisco Arts Commission for a permanent installation at the Randall Museum, is a wind-driven kinetic facade that transforms a blank wall into an observational instrument and reveals the complex interactions between wind and environment. Over the years, Charles Sowers has become increasingly interested in rooting his works in the dynamics and phenomena of their particular location. Many of these works are a blend of aesthetics and information. This has led to a kind of aesthetic/scientific instrumentation that reacts to a site and promotes insight into normally invisible or unnoticed phenomena. Through such work Sowers hopes to engage people in an unexpected dialogue with their locale and provoke the desire to notice the beauty and intrigue of the world around them.

The Randall Museum site, like many in San Francisco, is characterized to a great extent by its relationship to the wind. Climatically, offshore winds bring warm weather from California's Central Valley, while onshore winds produce San Francisco's famously chilly weather. Sowers knew he wanted to work with the wind. Initially he proposed an altogether different idea in a different location, but then the museum's staff articulated the desire to address a blank wall thirty feet high

Facade in motion
as viewed from
parking lot

(9.1 m) and thirty-five feet (10.6 m) wide, that faces the parking lot and is the first impression visitors have of their facility.

Drawing attention to the unnoticed or unseen is a dominant theme in Sowers's work, and *Windswept* is no exception. It presents actual physical phenomena that draw people into a conscientious noticing and interaction. The design consists of 612 freely rotating directional arrows, which serve as discrete data points indicating the direction of local flow within the larger phenomenon. Wind gusts, rippling and swirling through the sculpture, visually reveal the complex and ever-changing way the wind interacts with the building and its environment. The arrows are balanced so that they remain at rest in their last position; thus they preserve a snapshot of the last wind gust even when calm. In this way the piece conveys movement even when it is not actually in motion. The aim was to provoke a sense of delight and wonder and reward extended observation. Typically of Sowers, this work involves setting up conditions for some other force to animate or complete, whether that completion is achieved by the interplay of some natural phenomenon, the interaction of viewers, or both.

Windswept was inspired by scientific diagrams called vector fields, which show the

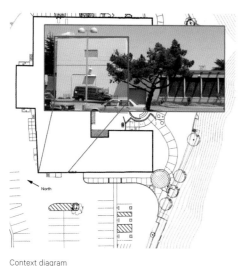

Context diagram

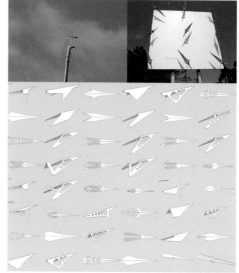

direction and intensity of certain phenomena, like wind, fluid flow, or magnetic fields. Sowers, who always found these diagrams intriguing, got the idea that he could make an animated physical analog. Like vector-field diagrams, *Windswept* seeks to reveal information about the wind. More than an informational graphing technique, however, it is a real-time, highly kinetic instrument that reveals the interaction between the site and the wind. Our ordinary experience of wind is as a solitary sample point of a very large invisible phenomenon; *Windswept* allows us to see the beautiful complexity of wind flowing through the built environment.

Sowers spent more than a year and a half prototyping and testing wind-vane designs, beginning by purchasing a number of commercially available maritime wind vanes, mounting them on a panel, and testing them in a wind. He quickly discovered that these products were inadequate to his requirements as they were not durable enough, and their bearings are designed for rotation in a horizontal plane. He needed them to rotate in a vertical plane as well as to be readable from a distance and from a vantage point other than that which maritime wind vanes

are made for. This led Sowers down a path of inquiry and experimentation with wind-vane design. His initial designs were rather naive, and his lack of understanding was quickly revealed when he held them out the car window while driving down the road or simply ran around with them like a six-year-old: the vanes would align perpendicularly to the airflow or oscillate back and forth. He made several dozen paper models of different designs and tested them at the beach or outside his apartment, in an investigation that was both functional and aesthetic. Ultimately Sowers tested eight different aluminum arrows, for nearly a year on-site and for a

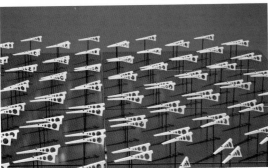

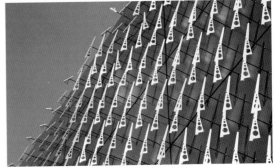

Details of facade in various states

year and a half outside his apartment window, in the very harsh salt air of Baker Beach in San Francisco's Presidio.

The final wind arrows are made of brake-formed anodized aluminum. The arrow axles are mounted to a standard, metal, architectural wall system consisting of twenty-five panels. The panel wall was set off the existing concrete masonry unity (CMU) wall to allow an equal volume of airflow for an HVAC vent that the sculpture covers. Holes were then punched in the panels in a twelve-by-twelve-foot (3.7 by 3.7 m) grid pattern, into which the installation contractor secured rivet nuts to accept the

stainless-steel axles. Once the panels were installed, the arrow assemblies were threaded into the rivet nuts.

The total installation time was just four days, and the result is has been gratifying for Sowers; he had understood how a small number of arrows moved in the wind but had been able only to imagine how the whole assemblage would behave; in this way *Windswept* truly has been an experimental instrument, allowing for the observation of a phenomenon that could be envisioned and possibly modeled but not persistently perceived in the real world.

REEF
Rob Ley, Rob Ley Studio and Joshua G. Stein, Radical Craft

Over the last century, the potential of an interactive, information-based architecture has shot off in varied directions. The most apparent scenario is often perceived to be that in which an individual is served by his or her surroundings. All of the precedent proposals rely on a degree of intelligence possessed by the machine. Essentially, the environment must be designed to be smart enough to predict—or at least understand in real time—the user's needs and wishes such that an immediate technological gratification quickly follows.

As fields outside architecture (transportation, medicine, communication, etc.) have evolved exponentially in proportion as they emanate from technological developments, architecture benefits from advances in material science, computational capacities, and information fluidity. Five years ago, Rob Ley Studio began a line of research grounded in the assumption that the responsibility of architecture can and should be more than providing shelter. Rather than continue with the belief that good contemporary architecture is that which serves, over the last several years the studio's work has focused on creating architecture that establishes a behavioral position. That is to say, an environment that serves the user is less beneficial than one that expresses a degree of willfulness and offers companionship.

Reef as viewed from Kenmare. The street interacts with the busy foot traffic moving between NoLita and SOHO.

The *Reef* project pulls from this research and comprises hundreds of responsive surfaces powered by nitinol, a shape-memory alloy, in its "muscle wire" form. This spatial arrangement, combined with an interface fed by an RGB camera and processed with software written for the Max/MSP platform, created an environment that responded to inhabitants. The processing of responses was guided by criteria ranging from basic location and proximity to more complex information such as the color of clothing and whether users were alone or in groups. More interesting than the arbitrary processing criteria was how the users responded once they realized

their presence and actions had an immediate effect on the space. Rather than focusing on the reasons for responsiveness, people were attracted to the nature of the motion, as in the dynamic between owners and their robotic pets. This consistent response by the variety of users has clarified earlier observations that intelligence and the capacity to process information may be overrated criteria within the study of interactive environments and artificial intelligence. Behavioral qualities—particularly kinematic motion and indirect responsiveness—are more successful in creating a connection between inhabitants and their environment. Continuing

Reef installation works in tandem with Acconci/Hall facade to create several interactive layers between private gallery and public sidewalk

research projects now look at the roles scale and materiality have in the dynamic.

Reef redefines the role of the architectural envelope by capitalizing on emerging material technology to imbue space with behavioral qualities. In this installation at Storefront for Art and Architecture in New York City, the public engaged in the new social nuances revealed by exploding the typical boundary separating private and public space. The exhibition's responsive membrane created a diverse range of porous and dynamic enclosures, capable of producing sophisticated, flexible interactions with an existing program. *Reef* produced an interior condition that reacted according to an exterior stereoscope and reasserted an active, willful role in shaping that public space. Although architects have often experimented with new technology, the discipline has been slow to investigate how material technologies can impact social structures within the built environment. The contemporary need for social overlap, idea sharing, and collaborative production demands that architecture develop alongside these emerging technologies.

Architecture's earlier flirtations with motion and technology have often been justified by claims of efficiency through intelligence. However, this territory of rational efficiency and intelligence quickly doomed architecture to the roles of the spectacular machine or the respectful servant. The heroics of Ron Herron's Walking City or even contemporary retractable stadium canopies rarely attempt to operate as a medium for social interaction. In these cases, technology drags with it a machine aesthetic, distancing it from the sphere of the social. Could a different paradigm expand the possibilities for viewing the human relationship with technology and space?

Reef investigated the role emerging material technology can play in the sensitive reprogramming of architectural and public space. For example, nitinol, was developed in the 1950s and '60s but has only been used in practical applications for the last twenty years. It is part of a class of metals called shape-memory alloys (SMAs), which are metals that change shape according to temperature. When these alloys are used as wires, a simple electric current offers enough heat to change the dimensions of the material, creating a wire that looks like any other typical wire but whose length can be precisely varied and controlled. Nitinol wire can contract along

Interior view

its axis by about 5 percent of its overall length, meaning that a length of 39 inches (99 cm) can shorten to 37 inches (94 cm). The effects of this small modulation in length can be amplified when the wire is used as an actuator for a larger surface. Mating SMA technology with a simple panel material creates a module that can alter its shape dramatically, offering the possibility of efficient, fluid movement without the mechanized motion of earlier technologies. Operating at a molecular level, this motion parallels that of plants and lower-level organisms that are considered responsive but not conscious. A field of sunflowers tracking the sun across the sky or a reef covered with sea anemones embodies the type of responsive motion this technology affords. Practical applications have been limited to the medical and aerospace fields as well as to novelty toys—the super exclusive versus the trite. Despite the potential of this technology, there have been few serious attempts to test its possibilities at the scale of architectural environments. *Reef*'s unique exploration of technology shifts from the biomimetic to the biokinetic while liberating and extending architecture's capacity to produce a sense of willfulness.

While much of contemporary architecture looks to nature for its form-making

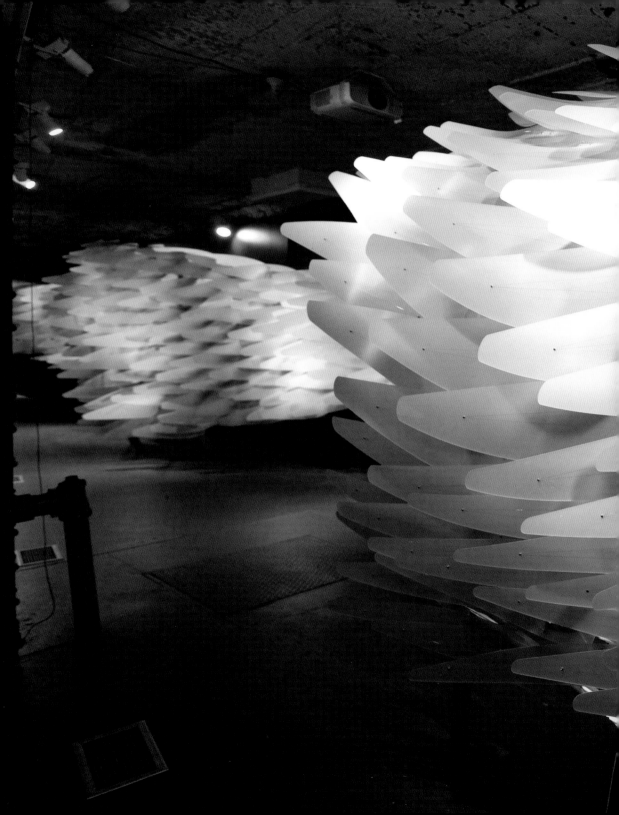

Aluminum tabs organize flow and orientation of fins and create an interface for dynamic control system

abilities—biomimetics—the impact is often limited to exactly this arena of formalism. *Reef*'s exploration of biokinetics allows the logic of nature to move beyond the merely formal, instead creating an environment of vigorous interaction. No longer looking to nature purely for static ideals of beauty, this project investigates nature's ability to form dynamic connections that start as physical but quickly lead to social implications. These lively relationships, so prevalent in natural ecosystems, seem to imbue nature with a consciousness that eludes rationalizations based on efficiency. Is it possible that technological implications in architecture

could be understood with this same sensibility? Stepping back from viewing architecture as an exclusively formal pursuit allows a more careful examination of its social effects.

Reef furthers the experimental agenda of Storefront through the investigation of a sophisticated and flexible negotiation of the public street and the typical first-floor retail space. The original building facade, by Vito Acconci and Steven Holl, engaged public space in a novel way by locating the art and architecture experiment at the interface between gallery and street rather than sealing it off from the public life of the street. *Reef* extended this experiment with the

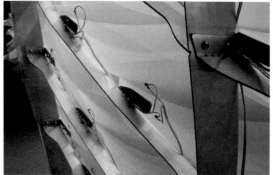

introduction of a more precise and fluid secondary interface, one charged with fostering refined social interactions through a variable and fluid porosity. Using an aggregation of six hundred responsive fins that form a singular surface, the installation negotiated the unique site condition of the gallery, imbuing the space with an identifiable personality while affecting the social patterns and movement patterns both inside and outside the gallery. Unlike the typical activities one associates with ground-floor spaces of New York City—retail, office, or gallery—here the motion and sway of nature, like trees in the wind, folded into the interior space, drawing in the sensibility of the outdoors. In tandem with the Acconci-Holl facade, the responsive surface was structured by an aluminum lattice capable of fitting snugly inside the specific context of the Storefront gallery. Within Storefront's planimetric wedge, this surface moved from simple vertical plane to volumetric habitable space, producing intense and varied experiential conditions. At the narrow end of the wedge, the vertical responsive surface acted as an interior billboard, drawing in the public from the busy street intersection. As the wedge of the Storefront's gallery expanded

from the corner, this surface wrapped back on itself to create a barrel vault of reactive fins, offering a more physically immersive experience. The surface was striated with a fin pattern running parallel to the storefront, creating a secondary layer of mediation inside the Acconci-Holl operable wall. The effect of the combined fin movement created local moments of visual transparency or opacity and altered the perceived scale and energy of the space, allowing it to selectively open visually to passersby.

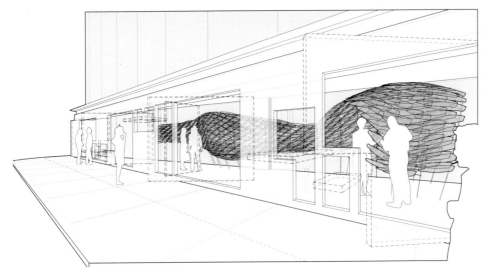

Perspective drawing showing animated social environment as it spills into the streetscape. An ambiguous, porous envelope created by both the building facade and the variable fin surface of the installation emboldens passersby to not only peek in, but to meander in from the public space of the sidewalk.

Diagram showing the installation as situated within the context of the gallery

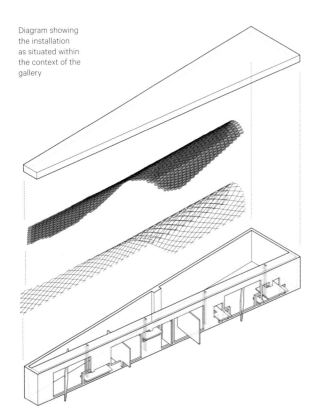

Animated sequence diagram demonstrates fin motion as it ripples through gallery space

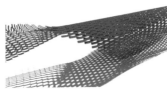

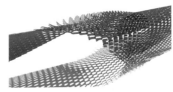

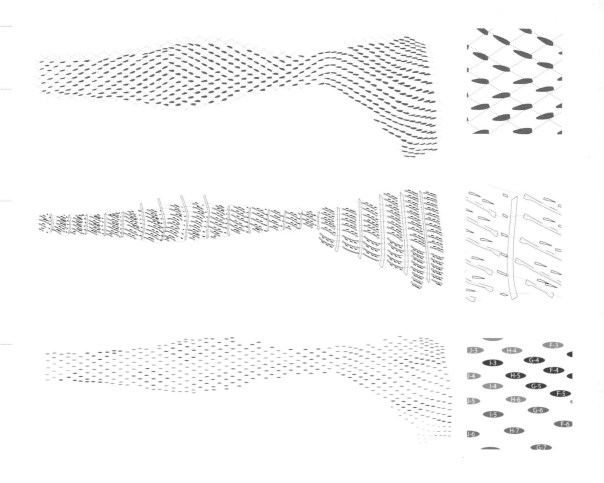

Assembly diagrams

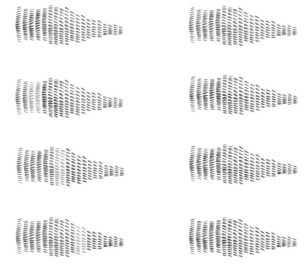

Illustration showing the arrangement of hundreds of muscle wire−driven composite fins attached to aluminum framework. Dashed curved lines indicate the range of motion exhibited by the kinetic fin elements.

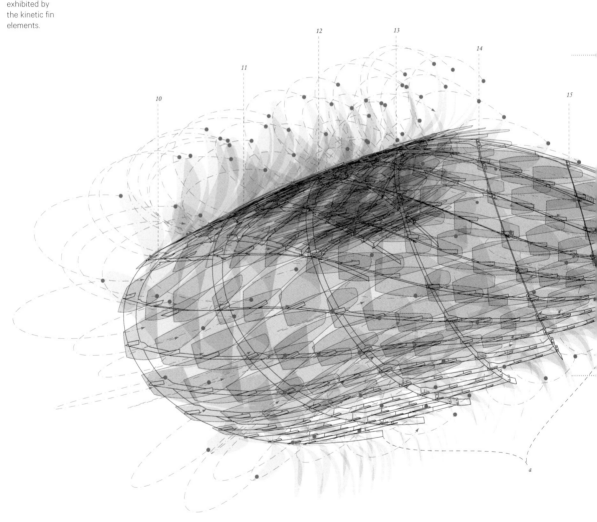

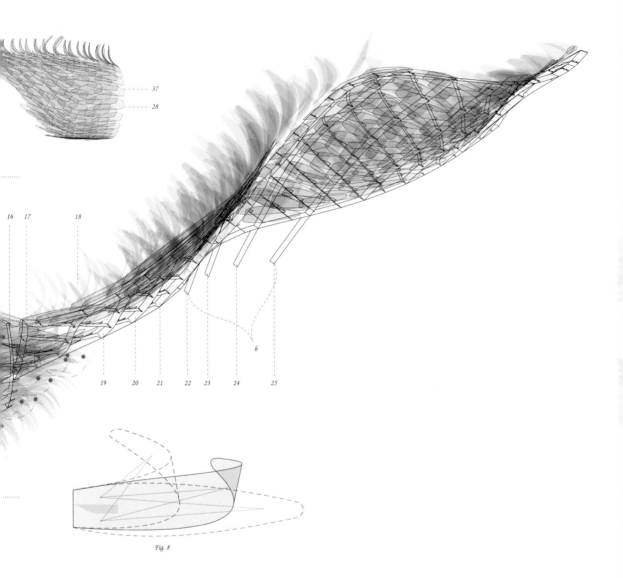

37
28

16 17   18

19 20 21 22 23 24   25

6

Fig. 8

COMMUN

Definition: to convey knowledge of or information about: make known

Related words: conduct, convey, give, impart, spread, transfer, transfuse, transmit

Merriam-Webster OnLine, s.v. "communicate," accessed March 25, 2015, http://www.merriam-webster.com/dictionary/communicate.

CATE

As our buildings begin to do more than respond, becoming truly interactive in the sense of facilitating conversations, we enter into relatively unknown territory with respect to our behavioral awareness. Our buildings are furthermore no longer limited to one-on-one conversations in that they can respond to, convey information about, and interact with assemblies of users. At the same time, such applications convey a level of interaction between the individual and groups of individuals. When architectural space has a true communicative capability, it fosters a heightened sense of attachment. As the theorist Nikos Angelos Salingaros states: "Our society tries to understand its own structure, and builds its physical extensions on the earth's surface, guided by the blank slate hypothesis."[1] The mind's reactions to form and environment have led to many mistakes in the past, as Steven Pinker points out: "City planners believed that people's taste for green space, for ornament, for people-watching, for cozy places for intimate social gatherings, were just social constructions. They were archaic historical artifacts that were getting in the way of the orderly design of cities, and should be ignored by planners designing optimal cities according to so-called

scientific principles."[2] It is important to recognize that people truly desire space, not style, and to understand the role that interactive architectural applications can play in facilitating such desires.

We do not inhabit architectural space simply for shelter; we do so because we need the experience of space, which hinges on a great number of factors at once, including lighting, acoustics, materiality, and the other people in that space. *Lightswarm,* by Future Cities Lab, combines many of these factors and communicates them to users simultaneously. *Lightswarm* also translates sound as the *May/September* project by Urbana does, but the Future Cities Lab project strategically applies the translation to a public lobby space. In this case, sound communicates an auditory pulse and amplifies it into cascades of light. Here the translation is that of a very perceptible phenomenon, one that simplifies the chaotic acoustic patterns of sound and engages the people in the lobby by making them appreciate and interact with it. Noise in public spaces such as lobbies and offices presents an interesting case of a phenomenon that can profoundly effect our experience of those spaces. Some lobbies, such as those designed for hotels, purposely manipulate the acoustic quality of the space to enhance the overall quality of the space. While this is typically done through the use of materials inherent to the space, its effect is often supplemented by the sound from ambient speakers. There are cars that use noise-cancellation technology such as that developed by Harman International Industries, but this kind of solution is not practical at the scale and complexity of buildings.[3]

Noise, after all, is often a positive factor in our understanding of space, how we behave in space, and even how we work in a space. Many studies have shown the effects of sound on our behaviors and cognition in general; a recent study out of the University of Chicago observed some direct correlations between sound and creativity. The study found that "ambient noise was the optimal level for creativity, whereas extreme quiet sharpens our focus, making it hard for us to think creatively."[4] While we may think it best to have an office environment with high levels of acoustic isolation, it may in fact be better simply to take the ambient noise, say, of cell-phone conversations and translate it into something more pleasing, like birds chirping or ocean waves. These noises help us stay acutely connected to the environment. The *Plinthos* pavilion, by MAB Architecture, takes a powerful approach to enhancing the sensory experience of the visitor, employing interactive mechanisms not as add-ons but as integral parts of the architectural materiality. They used clay brick because it is one of the most common and rudimentary materials that have human scale (a property inherent to its small modularity), and in contrast to the typical approach using contemporary materials, the brick is perceived as a symbol of solidity and reliability as the surface of the entire pavilion. The poetic combination of high tech and low tech creates a natural attachment to the space where the interaction takes place. High and low also play with the materiality of transparency not typically associated with brick, which allows for the manipulation of light, sound, and air.

The design of *BALLS!,* by Ruairi Glynn and Alma-nac, not only transforms the atrium of the engineering firm Arup's headquarters in a dramatic fashion, but also communicates a number of invisible data sets within the building as a whole, including, Glynn states, "current energy consumption within the building, the number of people working in the office, Internet traffic intensity, noise from meeting rooms, or even how much coffee is being consumed." Superficially, the balls simply change color as they rise and fall, but the important effect is that they can communicate real-time data sets about any number of phenomena within the building. There is no governing program for how the system behaves. *BALLS!* is essentially an instrument played via the behaviors of the participants in the building at any given time. Depending on the governing factors extracted from the data sets, the lobby can communicate an extremely wide palette of experiential qualities. An analogy would be the different sounds generated by playing the same drum with your hand, with drumsticks, or with wire brushes. The classic project *MegaFaces,* by Asif Khan Ltd., relies on communication at the most basic, literal level. The designers' self-stated interest was "to create a monument to celebrate people, regardless of their status as athletes or spectators, their age, nationality, sexuality, or gender." Almost the inverse of Ruairi Glynn and Alma-nac's project, which communicates conceptually complex data sets in a simple way, *MegaFaces* communicates a very simple concept—making an emotional connection with individuals—via complex technical means. The result is worth it, however, as over 150,000 people participated in its first incarnation at the 2014 Olympics in Sochi, Russia. The complexity of 3-D facial scanning, translated through the use of more than eleven thousand actuators, creates a monumental effect whereby the beauty emerges from the simplicity of the message.

LIGHTSWARM
Future Cities Lab

Lightswarm is a semipermanent interactive light installation in the public lobby of the Yerba Buena Center for the Arts (YBCA), in downtown San Francisco. The installation, measuring approximately one hundred feet (30.5 m) wide by thirty-five feet (10.7 m) tall, is located in a one-foot (0.3 m) sliver of space just behind the south-facing facade of the Fumihiko Maki–designed building. The facade faces the Yerba Buena gardens, above the Moscone Convention Center, and sits opposite a busy urban cinema and several neighboring museums. The existing nonoperable facade creates a highly reflective visual and acoustic barrier between the inside

and the outside spaces. The YBCA commissioned the installation both to provide shade to the lobby and, more importantly, to create a public participatory interface for the center.

One of Future Cities Lab's initial observations was that the facade's large glass panes subtly pulsed in response to passing vehicles and loud music. As a way to visualize this energy, they prototyped vibration sensors with LED graphic displays and attached them to various locations along the facade. By adjusting the sensitivity of the sensors, they discovered that they could capture a range of acoustically generated vibration coming from both inside the lobby and

Detail of fins

outdoors. This included sounds as indistinct as conversations and bicycles traversing the garden sidewalks. They then attached a grid of vibration sensors to the facade, creating what they called their "city sensor": a large-scale sensing array that eventually allowed them to capture and visualize a multitude of sounds and that created a real-time dialogue between the inside and the outside of the building.

As a part of the prototyping process, the Lab remapped and superimposed the data coming from the city sensor inputs over an elevation of the facade. Using Arduino hardware with Grasshopper and Firefly software, they graphically visualized this data in real time, creating a feedback loop between their variable inputs and a range of potential outputs, including light levels. This allowed them to explore the facade as a live field condition, populated with arrays of responsive LED modules that could geometrically transform through rotation, scale, opacity, and gradient changes. Additionally, they began to investigate the use of autonomous swarming algorithms that could amplify these effects and generate a range of emergent behaviors. Rather than an immediate local response, where sensor values and LED values are matched one to one, the algorithm

Interior atrium space showing vertical relationships

compares all incoming sensor values across the facade, identifies the largest value, and accelerates a virtual attractor point in its direction. Each suspended light module's brightness then increases as the virtual swarm passes by it. *Lightswarm* continuously seeks to spotlight the area of the facade with the most acoustic vibration, be it inside or outside, created by human or machine. Regardless of whether people are deliberately interacting with the facade or not, the algorithm always self-calibrates to pursue even the faintest sound. In this way, the facade revealed that it was in a state of continuous negotiation of the acoustic subtleties from the local site, the adjacent neighborhood, and the city beyond.

The final *Lightswarm* installation is made of 450 illuminated light modules within custom suspended frames. Each module consists of an intricately laser-cut hand-sewn shell that supports a 3-D printed chassis, which houses a microcontroller and LED module. Custom vibration "sensing spiders" are attached directly to the glass panes with clear adhesive. Running in the background, a Processing sketch receives real-time data from these sensors and uses it to drive a real-time swarm simulation. This simulation is then used to send out packets of

Views of interactive change over time

data to control the brightness of the individually addressable LED modules.

The result is a dynamic cascade of light streaming playfully across the facade night and day. At one moment its behavior might seem erratic and self-motivated; but upon deliberate human interaction (a tap on the glass, for instance), its behavior transforms into a swarming frenzy of local illumination. It is this transformation from passive to active that gives *Lightswarm* its distinctly eager-to-please—some might say neurotic—personality.

Diagram of facade connections

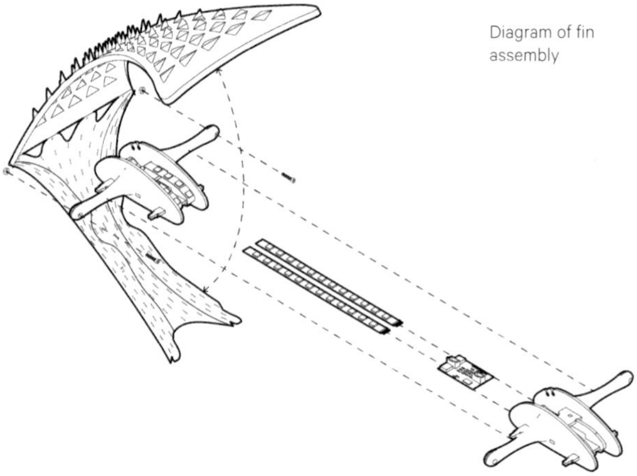

Diagram of fin assembly

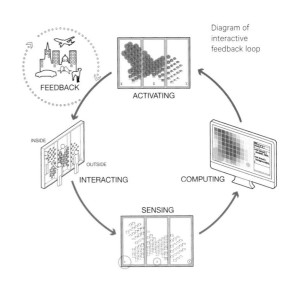

Diagram of
interactive
feedback loop

FEEDBACK

ACTIVATING

COMPUTING

INSIDE

OUTSIDE

INTERACTING

SENSING

Diagram of inside
and outside
interactive
relationships

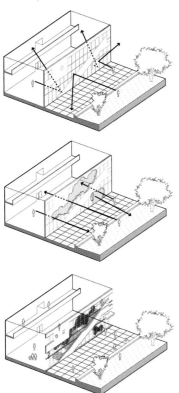

Sectional diagram
of human
relationships

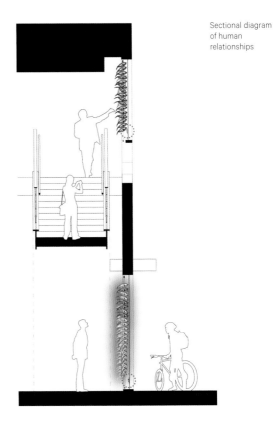

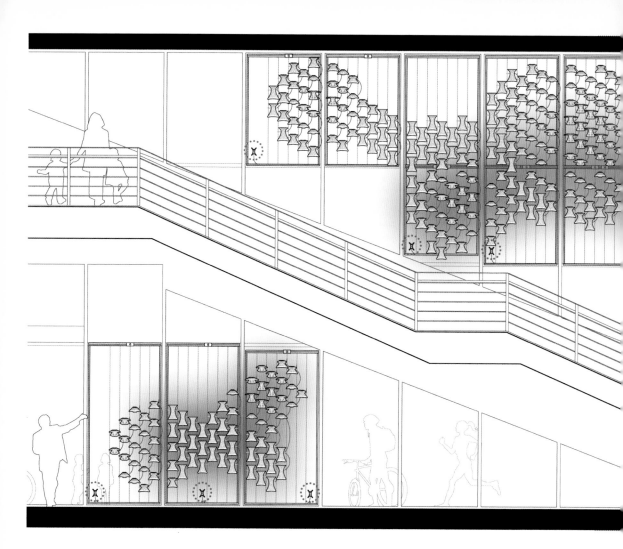

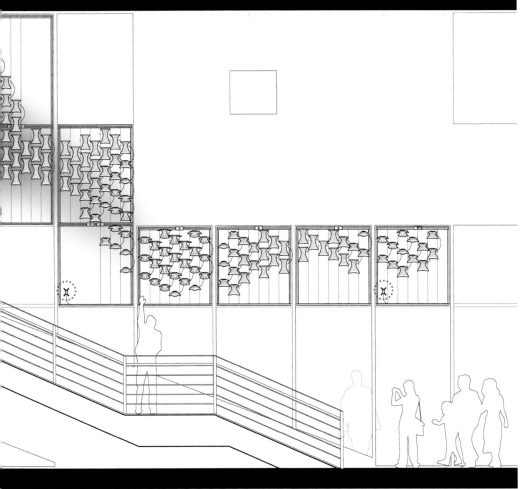

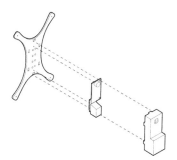

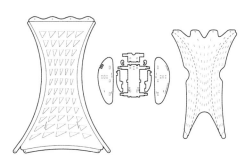

Axonometric
projections of
"sensing spiders"
and fins

PLINTHOS
MAB Architecture

The *Plinthos* pavilion was created for the ID10 Interior Design Show in Greece. MAB Architecture was commissioned by the event organizer to design a temporary pavilion from materials provided by sponsoring companies, one of which was a clay-brick manufacturer.

MAB Architecture was interested in using clay brick (*plinthos* in Greek) because it is one of the most common, rudimentary materials. It has been universally used as a building material for millennia, helping to shape the largest part of our built environment. Clay brick has human scale and is perceived as a symbol of solidity and reliability. Its warm color palette and cumulative surface texture often provide sensual qualities. There are countless shapes and sizes of brick, from traditional to contemporary hollow types; and with new technologies there seems to be no limit to the variety of brick bonds or the ways they can be stacked. With *Plinthos*, MAB Architecture wanted to explore these characteristics and possibly add another quality, transparency: something that is not normally associated with clay brick. To enhance the sensory experience of the pavilion's visitors, interactive mechanisms were employed not as an add-on but as an integral part of the structural design.

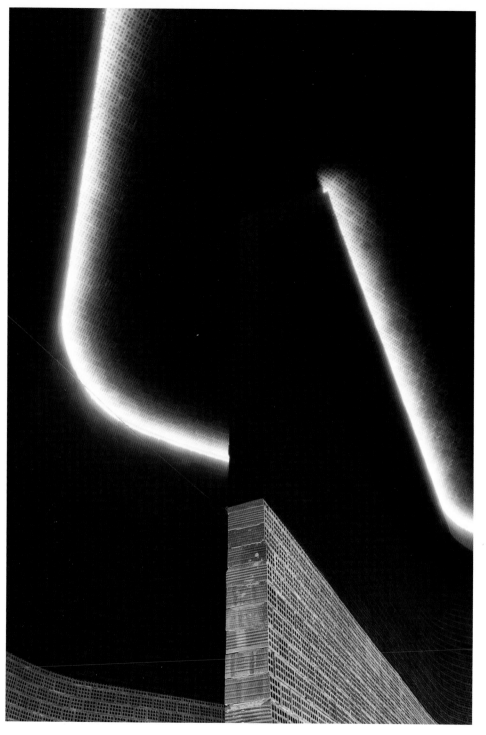

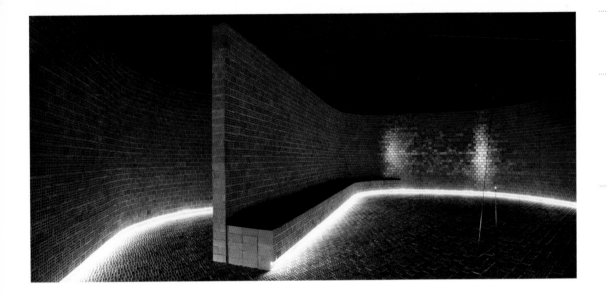

More than twenty thousand hollow bricks were used to complete the floor plate and walls. The usually hidden, perforated sides of the bricks were exposed, creating a permeable surface that allowed for the flow of air, light, and sound. The visual transparency created by the perforated brick wall became the channel of interaction between the visitors and the structure. Apart from the clay brick, the only other material used was a glossy, black, stretch ceiling membrane, which acts as a mirror and creates the optical illusion of a larger space.

The structure was built inside an exhibition hall, within a black box with a single entry point. The space between the outer skin of the black box and the inner brick skin was a service area, where most of the interactive mechanisms were positioned. The pavilion did not offer an external facade, as it was enclosed within the black box, but instead offered a theatrical entrance, which revealed very little of the interior. The entry sequence was carefully choreographed as the visitor moved from the brightly lit exterior of the exhibition hall to a dimmed, monochromatic, all-brick interior. The interior space breathed through its permeable wall and floor. A constant background soundscape and an expanding RGB light communicated through the wall, transforming the structure into a living organism in which the visitor gets completely involved.

Since *Plinthos* was built indoors, there was no need for any weather-protection measures. However, seismic activity is a frequent phenomenon in Greece; therefore, all the relevant structural and safety requirements had to be considered. Exposed brickwork is not common in this region, and local bricklayers traditionally work within a great amount of tolerance, because walls are normally plastered following assembly. It proved difficult to find bricklayers with the required skill and precision. The entire construction had to be completed within ten days, so efficiency and good coordination were a prerequisite. Eventually, a team of tilers was hired to build a brick wall with a satisfactory surface finish.

The construction began with the laying of the brick slab. This floor plate, measuring forty feet (12.2 m) by twenty-seven feet (8.2 m), became the carpet on which the wall of the pavilion sat. The wall, at a height of ten feet (3.1 meters), unraveled like a ribbon to create the

enclosure. Bricks were laid with minimal grout,
joined together by fast-setting tile glue instead
of the usual mortar. Structurally, the walls and
an approximately two-foot (0.6 m) band of the
floor brick slab on either side of the wall were
designed to act as a single element—an upside-
down *T,* shaped by bricks held together by the
tile glue. The rest of the flooring was formed by
loose-laid brick infill. The structural issue of the
stretch ceiling's lateral force on the walls was
solved by adding a perimeter steel-plate ring to
the top of the wall.

The *Plinthos* pavilion became a space of
exploration and contemplation—a sanctuary for
the senses—that allowed visitors to experience
architecture through an unusual set of lenses.
Its fluid design invited visitors to navigate within
thousands of perforations and form an intimate
relationship through masterfully articulated poetic
expressions. Together with the hard surface of the
structure, multiple ubiquitous layers of media and
other technologies existed and intended to blend
harmoniously into a performative/responsive
architectural installation. Each layer was responsible for a specific outcome; however, all elements
intercommunicated when necessary to express an
orchestrated intelligence as one of the construction's fundamental extensions.

Behind the walls of the vertically rotated
stacks of bricks, a big roll of reflective white
paper wrapped the environment in a protective
membrane, its main intention not to provide
safety but to reflect light emitted from a series of
interconnected LED bars positioned at the bottom of the exterior wall. The light fixtures were
tilted toward the white paper rather than the

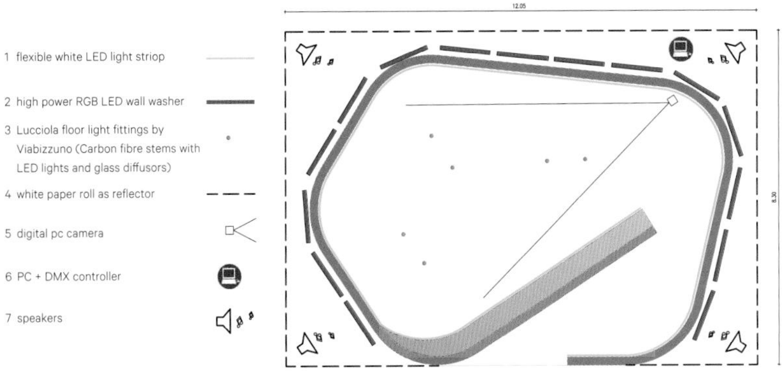
1 flexible white LED light striop
2 high power RGB LED wall washer
3 Lucciola floor light fittings by Viabizzuno (Carbon fibre stems with LED lights and glass diffusors)
4 white paper roll as reflector
5 digital pc camera
6 PC + DMX controller
7 speakers
12.05
8.30

Diagram of interactive systems

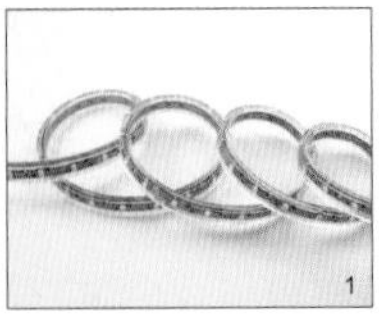
1
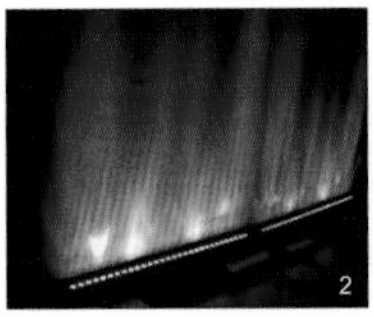
2
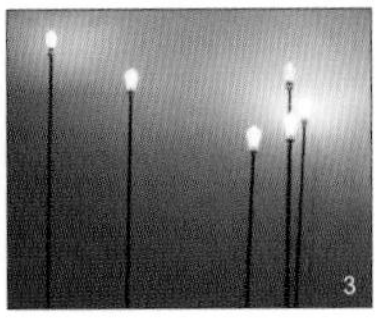
3
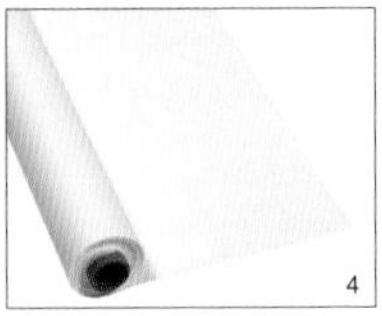
4

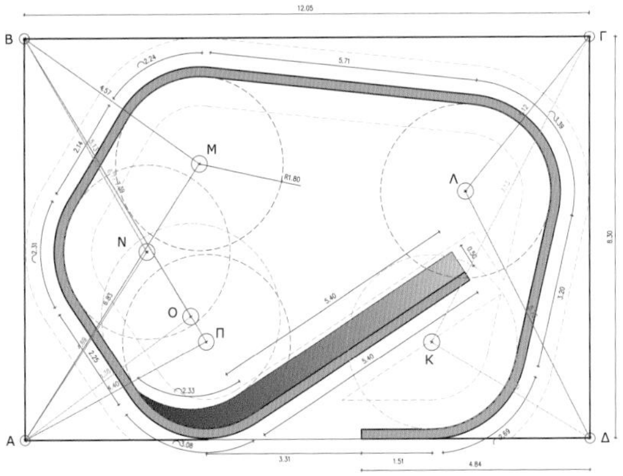
12.05
B
Γ
A
Δ
M
N
O
Π
Λ
K
5.71
8.30
3.31
1.51
4.84

Floorplan of environment

walls in order to reflect a uniform lighting coating that could be viewed from the interior through the transparency of the bricks' open cells. The light compositions were never in stasis; rather they oscillated through never-ending patterns to accompany real-time events taking place within the structure's interior. A computer vision-analysis algorithm was developed to scan and analyze a number of visitor interactions, such as motion, presence, position, circulation, energy, or formation clusters. The algorithm produced multidimensional information about spatial occurrences, helping the system identify patterns and drive the lighting compositions accordingly.

To create a synesthetic experience, real-time sonic compositions accompanied the lighting effects to enhance the feeling of immersion and to extend perceptual levels by way of aural excitation mechanisms. A large database of prerecorded and synthesized sound files were fed into the main processing system, which composes in real time the soundtrack of the space. Different layers of the sonic composition were generated from the sound files in combination with different processing algorithms to provide rich spectra of sonic images. The final result was projected through a multichannel speaker configuration that surrounded the environment, while all the

above: The pavilion
under construction
right: View of
pavilion entrance

Concept image of
brick translucency

audio equipment remained hidden beneath the surface of the hard structure.

Beyond the atmospheric background that became an extension of the spatial interactions, the stem lights of the interior space were used as triggering mechanisms for the reproduction of audio events. These reproductions were then added to the composition for a higher level of performative synthesis. Specialized flex sensors were embedded within the lighting objects so as to allow the system to perceive changes with haptics and object movement. The amount of motion defined the aggressiveness of the sound pressure and the processed texture applied.

The number of triggered light movements determined the selection of specific database categories, representing a level of complexity analogous to the recorded material that was created during the construction of *Plinthos*. That material included a great number of brick-working sounds, such as hits, smashes, scrapes, and breaks. The visitor listened to and experienced these events as snapshots of the structure's embryonic state and explored and reacted to the space as it reacted to the visitors' engagement and behavior. And *Plinthos* integrated yet another ubiquitous layer into its technological capabilities. Used as a medium to distort,

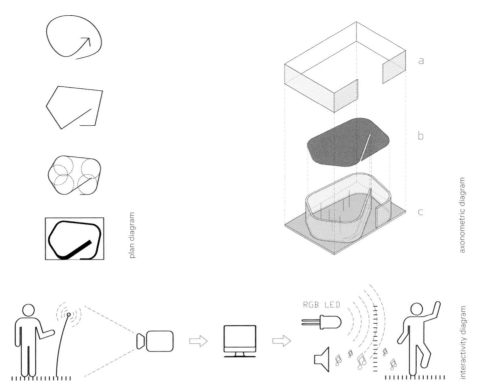

plan diagram

axonometric diagram

a

b

c

RGB LED

interactivity diagram

enhance, or provoke perception and communication, air fans positioned at specific hidden locations on the exterior of the environment were automated using proximity sensors.

Reflecting on the aforementioned technological abilities of *Plinthos*, one could easily assume that this installation became a manifestation of multiple disciplinary practices, synthesized through a high degree of sophistication for the sake of producing a memorable experience. Media and technology became fundamental criteria in the design of a sensate space, in this case allowing the physical structure to interact at numerous levels with the human subjects.

Thus, *Plinthos* speculated on the possibilities for an architecture consisting of layers of sensory perception and provided a unique meditative and affective environment for the provocation of feelings, senses, and desires. *Plinthos* bound the tactile emotion of the rough brick surface, a memory of ancient construction, to new technologies that stimulate the sense of time's passing, in short relating the past to the future.

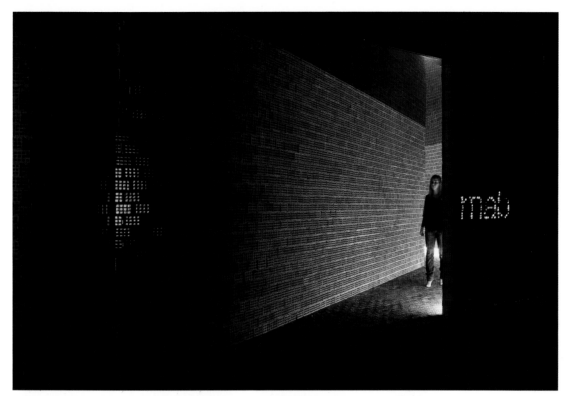

above: Pavilion entrance at night
left: Behind-the-scenes lighting

BALLS!
Ruairi Glynn and Alma-nac

Delivered in response to the theme of No.8@ Arup's 2014 competition, *BALLS!* is a data-responsive intervention in the dynamic relationship of control between buildings and their inhabitants. It presents an open-source platform, which allows inhabitants of the building, both as individuals and groups, to program and control the installation. The project allows inhabitants of the office environment to take more control of their building's behavior and stimulate unexpected interactions and creative outcomes.

BALLS! is an array of forty-two suspended glowing spheres in a six-by-seven-foot (1.8 by 2.1 m) grid that can individually rise and fall at varying speeds within the six-story atrium. Each ball is manipulated by its own robotic winches, all controlled by an open-source bespoke software that can convert input data into movement, color, and form. The winches run on a wired DMX system while the LEDs within the balls are controlled via wireless DMX. The winches needed to be customized to allow enough vertical movement for them to be seen from any floor. They were pre-fixed to trusses on-site, tested, given a DMX address, then moved into position. The whole installation and programming was completed within forty-eight hours, as specified by the competition brief.

Architectural
drawing of *BALLS!*
in context

The design team, which included six Arup departments, was led by the rigging company. For the first half of the installation, it programmed the installation to showcase its capability and begin a conversation about the relationship between buildings and their users. The intention was to show three modes of interaction: curated, where the installation is unaffected by occupants' behavior; passive, where occupants passively affect the installation; and active, where occupants actively engage with the installation. A microphone was set up in the atrium alongside contact microphones on the balustrades. The balls were programmed to move around a sine wave, which increased in amplitude with the noise level (passive interaction). The color was set to respond to the time (curated interaction), moving from blue to red over the course of a day. The contact microphones responded to the balustrade being tapped (active interaction), sending a ripple of white light through all the balls. For the latter fortnight and following a series of workshops, curation of the installation was handed over to the building's users, allowing the project to evolve into a more direct feedback loop between the inhabitants and their environment.

BALLS! is driven by enabling an open platform for behavior development. The code driving

the behavior of *BALLS!* was made available to all the inhabitants of Arup headquarters. Built upon the open-source programming language Processing, access to the tools required to manipulate and test code could be downloaded for free along with detailed documentation.[5]

Lunchtime workshop sessions were held for Arup's staff and the design team to develop novel applications for *BALLS!* The staff, primarily made up of engineers, was a captive audience for the project's technical appeal, but many participants came from administrative and facilities positions. Students of the Interactive Architecture Lab at the Bartlett School of Architecture supported

those not able to code. These workshops allowed people to feed their ideas to the design team and take an active role in modifying and improving the behavior of the system to reflect their daily experiences of it. Described in brief below are a few examples wherein codes were produced that eventually served as data for the behavior.

Puppeteer: Using a Leap Motion sensor, which tracks hands and fingers held above it, people could take control of the shape and height of the balls intuitively with simple hand gestures. An uncanny sensing of transmission, translation, and amplification was felt as small movements were projected onto the five-story installation.

Views of
the overall
environment
installed in
the atrium of
Arup's office
building

Sensor Grid: Combining with the Smart Citizen open-source platform, environmental sensor data from within Arup's building was recorded and transmitted into the atrium as an enormous 3-D scatter graph, allowing inhabitants to see how the data changed over the course of a day.

World Cup: Installed over the course of the 2014 World Cup, one particularly obsessive football fan chose to represent the flags of the nations participating in line with the schedule of matches.

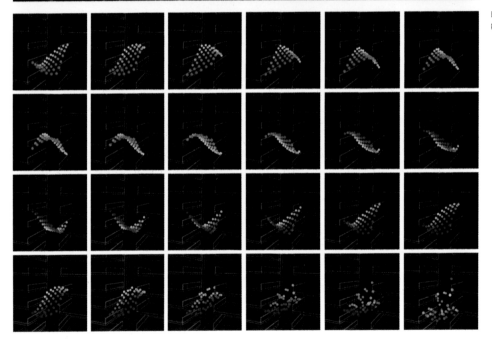

Bespoke software for controlling data flows

Diagram of variable possibilities

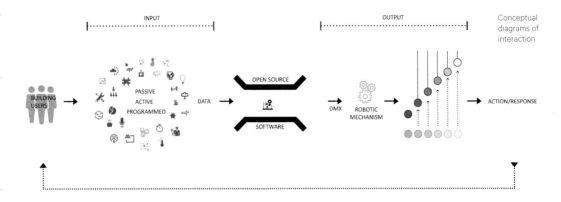

INPUT

OUTPUT

BUILDING
USERS

PASSIVE
ACTIVE
PROGRAMMED

DATA

OPEN SOURCE

SOFTWARE

DMX

ROBOTIC
MECHANISM

ACTION/RESPONSE

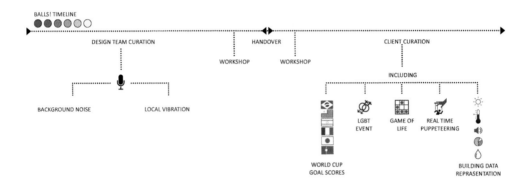

BALLS! TIMELINE

DESIGN TEAM CURATION

HANDOVER

CLIENT CURATION

WORKSHOP

WORKSHOP

INCLUDING

BACKGROUND NOISE

LOCAL VIBRATION

LGBT
EVENT

GAME OF
LIFE

REAL TIME
PUPPETEERING

WORLD CUP
GOAL SCORES

BUILDING DATA
REPRASENTATION

Conceptual
rendering of
balls in the work
environment

The installation
under construction

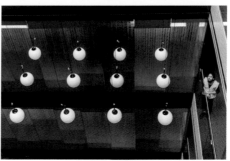

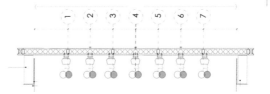

Diagrams of
installation
components

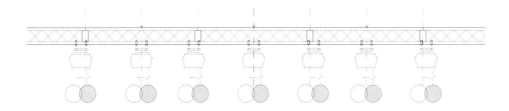

Exterior view of
atrium

MEGAFACES
Asif Khan Ltd.

MegaFaces was an experimental architectural installation. It comprised a large-scale kinetic-volumetric LED display, supported by a bank of automated 3-D scanning photo booths, an automated 3-D scan meshing system, a tablet app that used QR-code cards, an SMS notification system, an automated 3-D modeling and lighting algorithm, a Web portal, a video-streaming service, and a pavilion at the Sochi Olympic Park, which contained it all and hosted the 150,000 people that participated in its first incarnation. It was commissioned through an invitation-only architectural competition by MegaFon, one of Russia's largest mobile-telecom networks and general partner of the Sochi Olympics.

The client wanted to make an emotional connection with its 68 million subscribers, the Russian public, and the global audience of the Olympics. The practice's design was a building that could physically transform to take on the appearance of the people visiting it: a Mount Rushmore for the digital age. Each element of the installation was conceived, designed, and developed specifically for this project. The LED volumetric-kinetic display is a world first. The project was completed 377 days after conception.

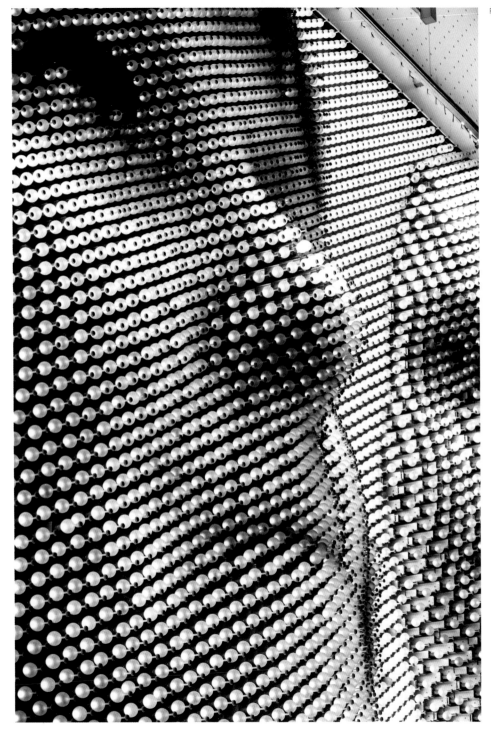

View of installation
in the context of
the public plaza

For thousands of years, people have used portraiture to record their history on the landscape and buildings and through public art. We are a long way from the days of Mount Rushmore and the epic figurative sculptures of Soviet artists such as Vera Mukhina, yet these monumental depictions of our heroes—whether presidents or the aspirational everyman—remain captivating to us. Today the Internet is how we record our history. Emoticons, selfies, Facebook, and FaceTime are just a few of the universal tools for communicating, and the face persists as the prevalent shorthand for emotion in these new mediums. Asif Khan Ltd.'s interest was in creating a monument to celebrate people regardless of their status as athletes or spectators, or their age, nationality, sexuality, or gender.

Formed by eleven thousand actuators, the volumetric-kinetic facade of *MegaFaces* transformed in three dimensions to re-create visitors' faces on a monumental scale. Facial impressions are relayed to the facade from proprietary instant 3-D photo booths, which traveled across twelve Russian cities before arriving at the Olympic Games and getting installed within the building. An electronic queuing system manages the face

data using individual QR-code cards, issued to each participant through a registration app, and enables participants' names to be displayed within and in front of the pavilion on screens, which also indicate the exact time the visitor's face will be appearing. The system also sends an SMS message to participants, relaying this information as well as a permalink to a live stream of their respective avatars appearing on the building. This allowed scanned participants across Russia who were without tickets to be part of Olympic history; their avatars were present even if they were not able to be at the park in person. Their selfie videos were shared over two million times on social-networking sites. The resultant portraits appeared on the side of the building, three at a time, at eight meters tall—a magnification of 3,500 percent—larger, in fact, than the Statue of Liberty's face. More than 150,000 giant selfies, of people from 106 countries, were shown over the course of the Sochi 2014 Winter Olympic Games.

With *MegaFaces,* Asif Khan Ltd. effectively created the world's first three-dimensionally actuated large-scale LED screen. The kinetic facade measures 59 feet (18 m) wide by 26 feet

above and top: Details of facade

above and opposite: Views of facade showing resolution from a distance

(8 m) high and consists of eleven thousand telescopic actuators, each with 120 internal components, arranged in a trigonal grid. Each actuator carries at its tip a translucent sphere that is a high-power RGB LED lamp. The actuators are connected in a bidirectional system that makes it possible to control each one individually and at the same time report back to the system its exact position. Each actuator acts as one pixel within the entire facade and can be extended by up to eight feet (2.4 m) as part of a three-dimensional shape or change color dynamically as part of an image or video that is simultaneously displayed on the facade. To maximize facial

recognition of each portrait, a scaling and positioning algorithm was developed that transforms the faces on the fly by considering daylighting, scale, rotation, form, and additional color.

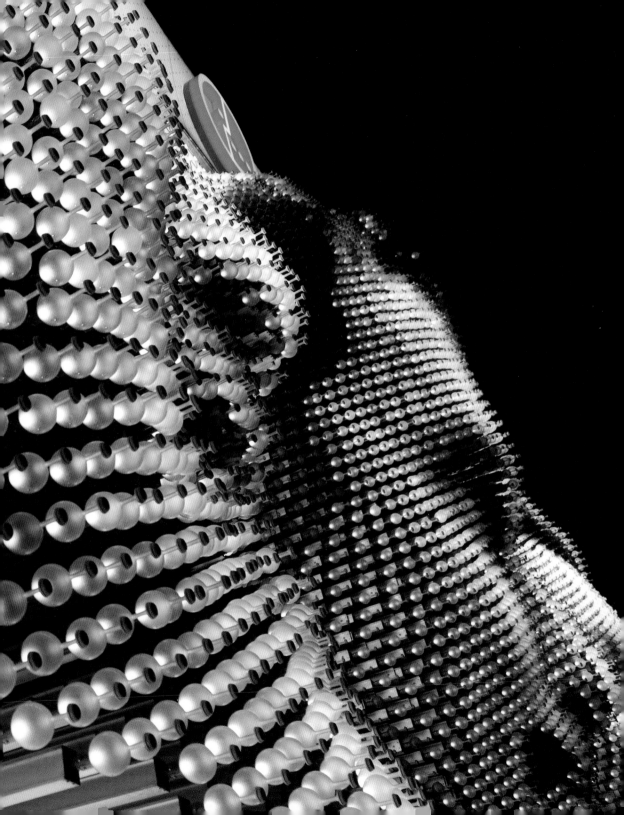

MEDIATE

[
Definition: acting through an intervening agency
Related words: equidistant; inmost, inner, betwixt and between, borderline, gray (also grey), in-between

Merriam-Webster OnLine, s.v. "mediate," accessed March 25, 2015,
http://www.merriam-webster.com/dictionary/mediate.
]

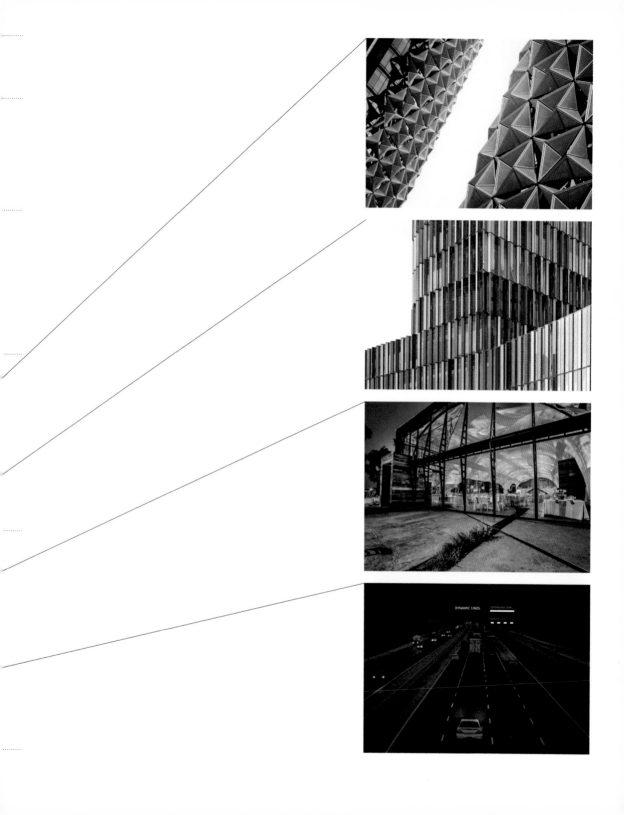

Mediation in this context refers
to the people within a building and the exterior
environment, both of which are in a state of constant
flux. The Roman architect Vitruvius, in his treatise
on architecture, *De architectura*, asserted that there
were three principles of good architecture: *firmitas*
(durability; it should stand up robustly and remain
in good condition); *utilitas* (utility; it should be
useful and function well for the people using it); and
venustas (beauty; it should delight people and raise
their spirits). Inherent to each of these principles
is the mediation of the people within the space
and the exterior environment. Of course, the idea
of a building facade that can mediate the changes
in human and environmental conditions is as old
as the first hinged window. Interestingly, prior to
the curtain wall and air-conditioning, most tall
buildings were not wider than thirty to forty feet (9
to 12 m), and they were well lit with daylight and
cross-ventilated. More often than not, however,
today's architectural solutions tend to shut out the
exterior environment completely and accommodate
humans by means of artificial lighting and heating
or cooling. Without air-conditioning, inoperable
windows and a glass curtain wall would render a
tall building uninhabitable. Anyone who has been
inside a car left to sit in the sun on a hot summer day
understands that without the ability to dump excess
heat, an internal environment quickly becomes
hotter than the outdoors. It should be noted that
operable windows on very tall buildings often do not
make practical sense, as wind speed increases with
altitude such that a small breeze on the ground is a
strong wind at the top of a building. Higher winds
combined with moisture can be very troublesome in
living units and particularly in office environments.
Furthermore, glass curtain walls are terrible
insulators, yet so often buildings are wrapped with
them in complete disregard for the sun and wind.

Thankfully, there has very recently been a
wealth of innovation and experimentation with
intelligent building facades. These facades have
been incorporated into the real world in a very
short period of time, primarily owing to the fact
that rethinking the building facade as a mediator
both saves money and makes better spaces. With
everything from kinetic panels that breathe and
titanium dioxide–covered walls that scrub the air
of pollutants to microalgae that work as bioreactors
inside panels, facades are clearly the single largest
area of innovation in adaptive buildings. Whole
books have indeed been published on the matter;
here we select a few pioneering projects from
the field. In the Al Bahar Towers, by Abdulmajid
Karanouh of Aedas Architects, we find the classic
glass curtain wall placed in the extreme desert of
Abu Dhabi. The innovative means by which this
project's mediating architectural boundary translates
into a dynamic, adaptive facade are truly an object
of delight. The faceted fiberglass rosettes, which
hinge on the cultural significance of the traditional
Islamic *mashrabiya*, are not placed all around the
buildings, but rather only where they are required
to be; and they preserve the views for the people
inside while dynamically blocking the direct glare
of light. The buildings' skins change throughout the

day, analogous to that of a living plant, adapting poetically to cultural and environmental forces.

The KfW Westarkade Tower, by Sauerbruch Hutton, takes a holistic approach to mediating the environment. The result is like an aircraft, where the necessary and appropriate response to all of the forces reconciles itself to a performalism of beauty. The overall form is streamlined to both integrate itself into its context and exploit the prevailing winds for natural ventilation. In the KfW Westarkade Tower, we see the building's skin and the structure as a whole responding to the urban, architectural, environmental, and human conditions. The high technology of the active building skin has a counterpoint in the atmospheric use of color, which adds the layer of chromatic variation as one moves along the street.

If we take a cue from Sauerbruch Hutton's holistic approach, we realize that there is great potential for adaptive architecture when one understands what a space currently does and how that space can aid in promoting or accommodating a specific change. This kind of spatial optimization can be defined pragmatically as a means for adjusting three-dimensional configurations according to the changing situations of both users and programmatic considerations. The development of a system that can accommodate spatial adaptability requires that optimization scenarios be analyzed both physically and organizationally. In architecture, kinetics implies relationships of cause and effect. To be able to design such a space requires an exploration of the dynamics of the users, the flexible possibilities for responding to those dynamics, and the adaptability of the architectural environment to undetermined changes. One way to begin is by rethinking architecture in terms other than those of conventional static and single-function spatial design, by emphasizing the dynamic configuration of physical space with respect to a range of constantly changing needs. *Eco-29* by FoxLin and Brahma Architects is a physically morphing, spatially adaptable wedding hall and event space that stands as one of the largest spatially kinetic environments of its kind. The architects' motivation for creating the design lies in mediating between the outside environment and the dynamic spatial layouts that exist sequentially in the context of a wedding ceremony. Natural analogies were studied to find a soft approach to the architecture, resulting in a concept that comprised a fabric skin kinetically morphing within the rigid confines of the building through the use of computer-controlled motors. It was required that the space be able to open up as much as possible for other events and that there be a stage at one end of the building and an open garden at the other. In effect, the space had to have absolute three-dimensional flexibility.

Smart Highway, by Studio Roosegaarde and Heijmans Infrastructure, takes a far different approach to a mediating interactive environment. This project's smart roads use light, energy, and road signs that interact with the traffic situation. Daan Roosegaarde is an artist, working in this case with an infrastructural company that shares his studio's goals of sustainability, safety, and perception; together they carried out several innovative applications of energy and light. In past discussions of smart highways, the focus has typically been on the car. Here the innovative point is that the roads themselves, not the cars, are what take on the intelligence. The highway is a work in progress, and the idea is that it repositions the role of designers in thinking about the possibilities of the environment itself, not just the objects in the environment, mediating architectural boundaries at a very large scale.

AL BAHAR TOWERS
Abdulmajid Karanouh, Aedas Architects

When in search of innovative designs for skyscrapers, nature and culture are the most persistent sources of inspiration. The Al Bahar Towers draw from both. The implementation of an advanced detection system was intended to integrate the building with its cultural context and respond directly to the needs of the region's climate: the *mashrabiya* screens, a type of wooden lattice shading, are anchored in the Islamic architectural tradition of the Middle East, and the dynamic movement of the individual units recalls the response of native plants to the movement of the sun. The architects used sophisticated techniques of parametric modeling and algorithms to facilitate their approach, which allowed them to refine the design concept via computer programs without sacrificing the ambition of the project.

Design teams always face the major challenge of how to realize a building that, within a constantly changing environment, provides its users the required comfort in an efficient and cost-effective manner. Design decision makers are caught between wanting to comply with green regulations and receiving the related certifications, on one hand, and implementing pragmatic and efficient solutions in a relatively standardized and rigid industry on the other.

Emerging from the landscape, the energy- and structural performance–driven design dictates the organic form and geometric patterns of the surrounding kinetic shading components

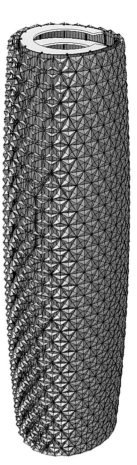

Conceptual drawing of tower with mashrabiya facade

Details of mashrabiya facade system

Traditional mashrabiya shading screen

Detailed units produced in Digital Project software by Gehry Technologies

Determining the best nonstatic, one-size-fits-all solution that will perform highly in a dynamic context is challenging indeed. In nature we find many solutions to similar problems; natural systems are anything but static as they constantly adjust and adapt to their changing environment from day to night and season to season. Natural systems were a major inspiration for the concept design of the case study presented here, which incorporates a novel, smart-adaptive solar screen that follows the movement of the sun.

Over the past two decades, naked glass has come to dominate the majority of the building facades of both general and iconic office towers, particularly in the Persian Gulf region. They have in turn become the norm in terms of clients' and users' expectations, sponsored and encouraged mainly by local governments. It will take some time, therefore, to reeducate the market and reorient expectations accordingly. The Aedas design team viewed the introduction of the dynamic mashrabiya screen as a step away from completely naked glass facades and toward rethinking this devotion to the transparent, which the authorities in Abu Dhabi are now showing encouraging signs of doing.

The design brief for the towers was based on the desire to create a building that would represent the ethos of the Abu Dhabi Investment Council (ADIC), housed in one of the buildings, while using a contemporary idiom to reflect the underlying cultural tradition of Abu Dhabi. Islamic and regional architecture, sustainable technology, and inspiration from nature form the triangular foundation of the design concept for the ADIC towers. For centuries, traditional Middle Eastern architecture has been known for its sustainable elements, including wind catchers, solar screens, cooling courtyards, ventilated domes, subtractive self-shading geometries (e.g., muqarnas), and many others. These features help provide comfortable spaces in extreme weather conditions and temperatures reaching up to 122 degrees Fahrenheit (50°C) and humidity of up to 100 percent. Carefully avoiding the mockery of traditional styles or direct mimicry of natural systems, the design aims to recapture some of the sustainable features used in the past and derives bio-inspired methodologies to enhance the buildings' performance. The power of the concept is in the computational algorithm, which merges the various design principles and translates them

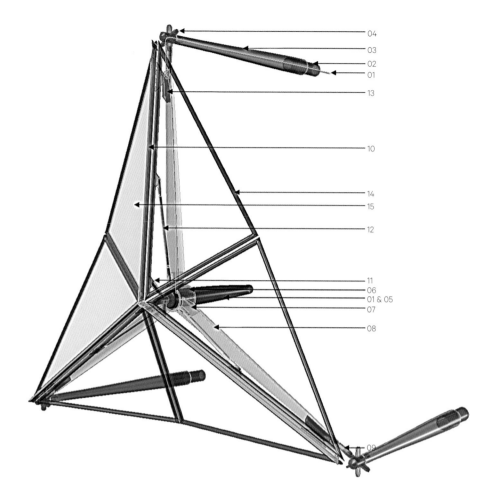

04
03
02
01
13
10
14
15
12
11
06
01 & 05
07
08
09

Dynamic Mashrabiya Unit Main Components

01   Actuator + Power & Control; cable connection back to the tower
02   Strut sleeves; penetrates the curtain wall & connects to the main structure
03   Supporting Cantilever Struts; hooks on the sleeves
04   Star pin-connection; receives the unitized Y-arms ends
05   Actuator Casing; protects the actuator
06   Y-Structure Ring-Hub; joins the Y-arms and actuator together
07   Y-Structure Sleeves; connect the Y-arms to the hub
08   Y-Structure Arms; support the whole mechanism
09   Y-Node Pin-Connection; pins to the star connection
10   Y-Mobile Tripod; drives and supports the fabric mesh frames
11   Actuator Head Pin-Connection; pin to the mobile tripod
12   Stabilizer; takes the loads to the hub releasing the actuator shear forces
13   Slider; allows the mobile tripod to travel along the Y-arms
14   Fabric Mesh Frame & Sub-Frame; supporting the fabric mesh
15   Fabric Mesh

Single unit in the
Scientific and
Technical Centre
for Building (CSTB)
wind tunnel in
Nantes, France

Single unit being
prepared for
durability cycling
test in Basel,
Switzerland

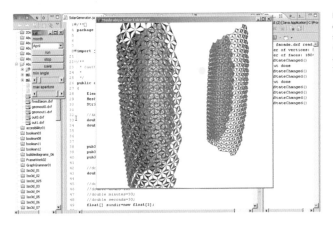

Bespoke Java control program, simulating the motion of the sun and the reaction of the shading screen

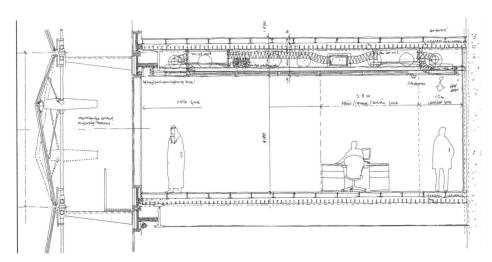

Section drawing of shading screen and tower interior

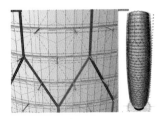

Sample diagrams from the 560-page Geometry Construction & Performance Manual developed for the towers

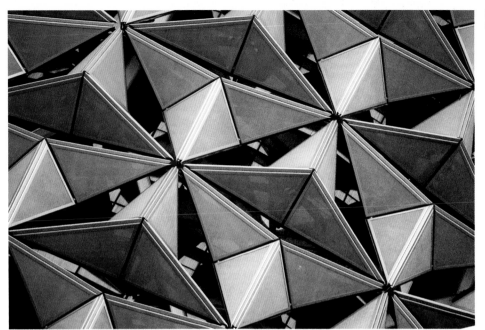

into a performance-oriented form and integrated mechanized system able to adjust and adapt to the changing environment.

The geometric composition of the buildings consists of the intersection of infinite arrangements and populations of circles (2-D) and spheres (3-D), which in turn generate infinite arrangements of nodes that, when connected, follow a mathematical logic that produces familiar Islamic patterns and forms. The floor plates and vertical profile of each tower are made of tangential arcs that also adhere to specially devised mathematical rules, resulting in an intelligent, fluid form that maximizes the volume-to-envelope ratio and natural-lighting distribution while minimizing wind-load impact on the building skin and infrastructure. The main structure is composed of a honeycomb formation, uniquely applied to the towers to provide a highly efficient and redundant structure. The envelope comprises a series of smart, folding kinetic shading components that act as a screening layer to the weathering glass skin, protecting the building from excessive solar gain.

A relatively clear glass curtain wall forms the weather-tight layer of the towers' skin. A secondary layer comprises intelligent, automated shading components—opening and closing via

The mashrabiya
units are designed
to respond to
weather changes
in the manner of a
planet reacting to
the environment

centrally located linear actuators—linked to a
computerized control system that follows the
sun's path. The shading screen acts as a dynamic
mashrabiya; it reduces solar gain and glare while
improving visibility by avoiding dark tinted glass
and blinds, which distort the surrounding view.
This intelligent system better distributes natural
diffused light, optimizes the use of artificial
lighting through dimmers linked to sensors, and
reduces air-cooling loads, ultimately helping
to reduce overall energy consumption, carbon
emission, and mechanical room size.

The dynamic mashrabiya is a unique kinetic
shading system that comprises triangular units
subdivided into six triangular flat frames that fold
like an umbrella at various angles, providing fins
and louverlike geometries in various directions
and positions. The design concept was inspired
by natural adaptive systems like leaves, flowers,
and skin spikes that open and close in response
to the environment—in particular, to the sun. The
flexible, smart folding geometry was carefully
worked out to overcome the limitations of traditional vertical and horizontal louvers, especially
as applied to geometrically complex buildings.
There are 1,049 units fitted to each of the towers,
covering the east, south, and west zones, leaving
exposed the north face, which has no exposure to

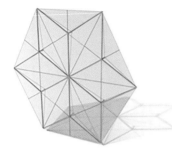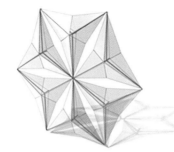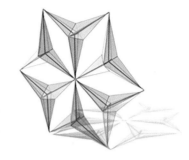

Mashrabiya units
at unfold, midfold,
and maximum-fold
positions

direct sunlight. When a facade zone is subjected to direct sunlight, the mashrabiya units in that zone deploy into their unfolded, or closed, state, providing shade to the inner glazing skin. As the sun moves around the building, each mashrabiya unit progressively opens.

The Y-structure arms support the entire mechanism and translate the forces from the supporting cantilever struts to the strut sleeves, which protrude from the curtain wall and back to the main structure of the towers. The mechanism is driven by a centrally located, electric screw-jack linear actuator designed to perform in aggressive, sandy environments. The actuator stroke reaches up to one thousand millimeters, folding the mechanism and providing up to 85 percent open area. Electrical outlets at each floor power the actuators and controllers. Data transmission from the control room. received by actuators, travels via a dedicated ethernet to PLC controllers using supervisory control and data acquisition (SCADA) protocols. To provide a tidy and simple approach, the routing of data and power cables to the actuator is concealed inside the struts and Y-structure arms. The Aedas research and development team initially collaborated to produce a bespoke program using a Java stand-alone applet, which

simulated the path of the sun and the kinematics and reaction of the shading units. A Siemens control system was utilized to program the actuators' stroking distances and operation-time durations. The key benefits of the responsive facade are the many advantages to the users of providing the following improvements in their working environment: better naturally lit spaces, better external natural views, fewer obstructive blinds, improved comfort by reducing air-conditioner flow and drag, and a unique, entertaining feature. In addition to these advantages, the mashrabiya system provides the following benefits to the whole building: 20 percent energy saving (up to 40 percent to the offices), a 20 percent reduction in carbon emission, a 15 percent reduction in plant capital cost, and a LEED Silver certification.

The dynamic mashrabiya solar screen of the Al Bahar Towers may represent a new benchmark in the field of adaptive building systems. Just as with other technologies, the more popular and common this system becomes, the more its reliability and affordability will increase. The design-to-delivery process described may also offer a roadmap for developing novel solutions that promote sustainability in a more genuine and effective manner.

KFW WESTARKADE TOWER
Sauerbruch Hutton

The KfW Westarkade is a prototypical ecological building and one of the greenest office structures in the world. It plays a pivotal role in the struggle against climate change and for conservation of resources—two of the most important social concerns of our time. The apparent ease of the aerodynamic form and its understated elegance belie the complexity of the building's strategy for resolving multiple urban conditions and implementing advanced environmental solutions.

Located in Frankfurt, Germany's West End and bordering the central Palmengarten Park, the Westarkade creates a unified composition out of KfW Banking Group's formerly disparate office buildings from the 1970s, '80s, and '90s. A new four-story low-rise clearly delineates the edge of Zeppelinallee, a much-traveled arterial road that leads north and provides the Westarkade with both entrance and address. A streamlined fourteen-story structure emerges from this low-rise, standing between the busy street and the cluster of towers from the 1970s, which are located at the center of the site and house the executive offices. The figure of the new high-rise is specifically designed both to maintain the views from the towers and allow them to remain visible from the street, while at

left and opposite:
Various views of
the exterior facade
with changing
perceptual
qualities

the same time accommodating a considerable building mass on the site. To the south, the new low-rise combines with the main building and the neighboring Nordarkade to form a communal courtyard. This extends the adjacent Palmengarten for a coherent open space that flows into the organically shaped foyer of the Westarkade and recharacterizes the existing buildings in the complex.

The KfW Westarkade is a tangible ecological example of how a high level of comfort can be achieved for some seven hundred employees in a thoroughly sustainable way while reducing energy expenditure to a minimum. In concrete terms, the building's annual primary energy demand is pared down to less than a hundred kilowatt-hours per square meter, which is considerably less than what is used by conventionally built office blocks. This goal was attained through a precise combination of measures that emerged from truly innovative planning and design as well as from the architectural and engineering teams' strictly interdisciplinary method—not to mention the consistently high aspirations of the client.

From the outside this ecological approach is epitomized by the aerodynamic form of the tower, oriented toward the predominant wind

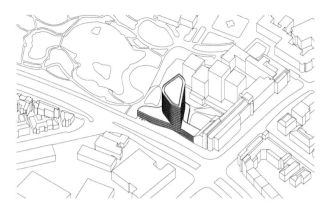

The outer layer wraps around the building with a zigzag profile, its wide areas of transparent glazing alternating with narrow ventilation flaps. Largely aligned to face the prevailing wind, these flaps automatically open and close according to external conditions, so that both temperature and pressure remain as consistent as possible in the interstitial space. On summer days, the facade can also be opened all around to prevent overheating, while integral blinds within the buffer space provide each office with individually controlled solar protection.

Using this system of protected opening windows, the offices are naturally ventilated for eight months of the year, with minimal dependence on outside conditions and without drafts or unwanted heat loss. When outside temperatures rise above 77 degrees Fahrenheit (25°C) or fall below 50 degrees Fahrenheit (10°C), a mechanical ventilation system supplies clean, fresh air, drawn in from the Palmengarten and pre-tempered in an underground geothermal heat-exchange duct before entering the offices. The deep floor plans of the high-rise were conceived to offer flexibility in configuration as well as comfortable and enjoyable office and circulation spaces. Single, double, and group offices all have generous views of the surroundings and the Frankfurt skyline; each is accessed from the curved and diversified spaces of the corridors. These corridors are designed with exterior windows for light, orientation, and view, but they also receive natural lighting via the floor-to-ceiling glass doors of the offices. The circulation cores are placed at the center of the plan and

above: Isometric drawing of tower and urban context

left: Performative qualities of the "pressure ring" in plan

direction and by the unique double-layered, wind-pressurized facade. This active facade has a precisely defined function in the climate concept of the whole structure, combining performative, technological, urban-spatial, tectonic, and atmospheric considerations into a single constructed form. The skin is made up of two layers: the double-glazed facade of the offices, with normal, operable windows, is enclosed within a second single layer of glazing and an air buffer between the two layers. This creates a "pressure ring" around the inner windows, protecting them from turbulent weather conditions that otherwise prevent their being opened at such heights.

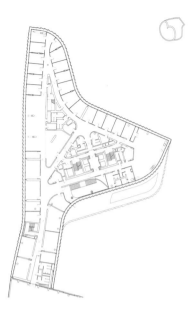

0   10   20m

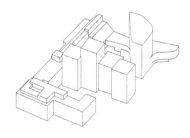

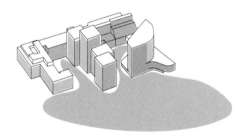

Overall massing
diagrams in
response to
neighborhood
context

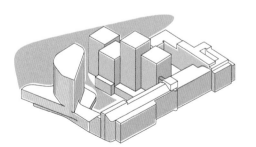

Elevation showing spring and fall ventilation

↑ mechanical ventilation of core areas

↳ natural ventilation via manually operated windows ventilation of core areas

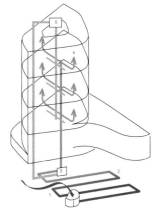

Elevation showing summer ventilation

1 incoming air from Palmengarten
2 geothermal heat exchanger
3 HVAC centre
4 incoming air through raised floor
5 heat recovery
6 exhaust air shaft using stratification

Elevation showing winter ventilation

1 incoming cold air from Palmengarten
2 geothermal heat exchanger
3 HVAC centre
4 incoming air through raised floor
5 heat recovery

Details of facade
system

terminate, where the corridor widens, in communal spaces such as meeting points, conference rooms, mailrooms, and small kitchens.

At street level, a pedestrian's every few steps bring a change in the building's appearance. Both the form of the building and the expression of its outer skin alter as the high-rise mediates between various city spaces: viewed from the intersection of Bockenheimer Landstrasse and Senkenberganlage, it appears as a tall, narrow tower creating a significant visible accent; from the north, along Zeppelinallee or Sophienstrasse, the wide, curved form creates an elegant counterpoint to the flow of traffic; and from the park areas of the Palmengarten, the building provides a discreet shimmering backdrop. So within a short journey, KfW Westarkade can be experienced in a number of ways: as a neutral, slender glass volume; as a convex taut skin; and as a wall of pure color. Color has been applied to the outer facade on the narrow ventilation flaps in three families, each of which addresses a different city space: green tones front the Palmengarten; red hues of the main sandstone prevalent in urban Frankfurt are interpreted along Zeppelinallee; and a group of blues and grays complement the more neutral scheme and materiality of the company's recently renovated central towers.

When viewed from the northeast, the building appears transparent and monochromatic, because the colored ventilation flaps are strictly oriented to the prevailing southwesterly winds. Conversely, standing on the street in front of the Westarkade and looking up at the southern edge of the high-rise, a myriad of reds and blues accelerate toward the upwardly thrusting point of the tower. The optical transformations that arise from the interplay of the Westarkade's streamlined form with its zigzag skin and chromatically varied facade are a direct expression of the building's sustainable energy concept. They are particularly visible on the wide, curved side of the high-rise, with its flowing changes of color, as well as on those parts of the building seen obliquely. Form, color, and material combine in the KfW Westarkade to bring diversity and stimulation to the urban space and enrich everyday city life.

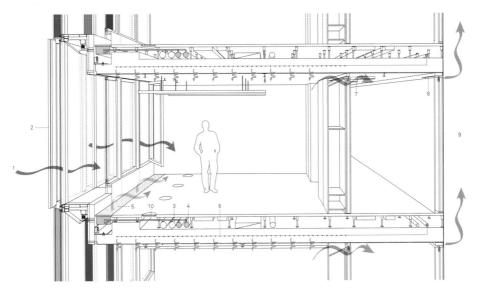

Section diagram of performative qualities of the building's office ventilation

1 fresh air supply
2 air vent
3 central heating lines
4 central cooling lines
5 air vents or underfloor convectors on every second axis
6 thermoactive building components
7 sound-attenuating air overflow
8 central exhaust
9 vertical exhaust shaft using natural stratification
10 underfloor electrical outlet

Architectural diagram of facade details

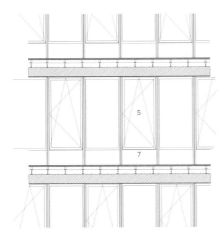
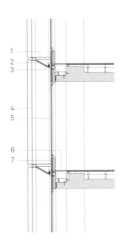
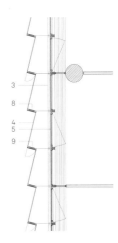

1 traversable aluminium sheeting above, steel sheeting for fire protection below
2 insulation covered by perforated aluminium sheeting
3 solar protection: blinds with light-redirection feature
4 secondary facade: single-sheet safety glazing
5 main facade: every second module is an operable window
6 underfloor convector or air exhaust
7 parapet: sandwich with exterior acoustic insulation and interior thermal insulation
8 fixed panel with acoustic insulation
9 ventilation panel on every second module to allow for summer cooling, interior acoustic insulation

ECO-29
FoxLin and Brahma Architects

Arguably, the most innovative designs utilizing kinetics arise from the need for site-specific, situational use, and it is this use that acts as a driving force in the changing and evolving patterns of human interaction with the built environment. Initially, *Eco-29* began as an exercise in creating a space that could change very rapidly to accommodate a variety of layouts and scenarios inherent to the context of a wedding ceremony. The notion quickly arose that the building, located in Israel, would be capable of physically encouraging hundreds of people to move around the space. After several weddings were attended and the sequencing of space throughout the

ceremony was established, the concept was then diagrammed to show how it could accommodate a number of spatial scenarios. For events such as a pop-up market or corporate product launch, the space would need to be opened up as much as possible, and there would have to be a stage at one end of the building and an open garden at the other. In effect, the space would have to have absolute three-dimensional flexibility.

The solution was to have mechanized walls and ceiling, and it was thus decided that cladding the entire space in a moveable fabric would provide the most effective means of realizing that solution. A number of conceptual renderings

Exterior and interior views of the kinetic space

were then developed in order to understand the aesthetic capabilities of the approach. There were many obvious advantages to using fabric, including projection, lighting, and the economics involved in having a mechanistically simple approach. Numerous physical models were created at various scales from fabric that had stretch properties proportional to what would be used in the actual project. Although it was very early on, a specific type of four-way stretch fabric was chosen, providing a valuable design constraint. Various schemes for attaching the fabric were devised, and it was decided that a series of vertical and horizontal ribs would be used.

At this point, it was decided that there would be a total of fifty motors, each rib having seven motors (five on the rib itself and two on the floor). The two ends of the system would be fixed and mounted to rolled steel tubing. Of the five motors on each rib, three would be clustered on one side and two on the other, each cluster positioned near the top of the building. The two floor motors would sit under the floor itself. Pulleys would have to be used for the top three attachment points in order to have the lines to the motors clear the fabric in a fully extended position.

Two connections were developed and tested for the fabric connections, which are not typically

left and opposite:
Views of
construction and
full-scale mock-up

Comparative
views showing the
dynamic flexibility
of the fabric

subjected to constant dynamic forces. The connection point to the fabric was a simple knot held by a sphere, developed so as not to pinch the fabric, and this became a standardized detail at every instance. The floor connections were a much more challenging task, as the ribs needed to telescope at the base and change their length up to ten feet (3 m) on each side without losing strength, and they needed to move in a line along the floor while having free rotation. Numerous designs and prototypes were developed in an effort to solve this detail; in the end a fixed screw-drive was used in the floor, connecting to a base for windsurfing boards. The rib was

allowed to telescope on this moving joint, which anchored the fabric.

There were two major issues in producing the fabric. The first was the aforementioned curvature, created when unrolling the geometry; the second was the fact that when the diameter of the rib increased, the line of the base in the track moved at a differing angle. In preparation for sewing the full fabric, FoxLin had to make several compromises based on the client's desire to have the fabric perform optimally at the most open position. In other words, the fabric could not sag or crease when fully expanded. It was thus calculated that the unstretched fabric

should be sewn at sixty-eight degrees, so that when it would be in the fully stretched position and the base attachments were near the outside of the building, they would hit the point on the track that is at forty-five degrees relative to the structure of the building. FoxLin also calculated a 10 percent pre-stretch of the fabric in the horizontal (long) dimension and added ten feet (3 m) of extra fabric on the short dimension for the telescoping base connection.

Two types of motors were used to address the different torque and speed requirements, and custom software was developed to choreograph the overall motion. The motors used were NEMA 42 high torque hybrid stepper motors with a six-ampere peak current per phase and a peak line pull of about three hundred pounds (136.1 kg) each. For the two motors that were used to pull the points near the center of the arch, a 5-1 worm gearbox and larger-size pulley were used. The geared motors used a larger pulley to approximate the speed of the nongeared motors. The nongeared motors had a peak line pull of about 150 pounds (68 kg) with a smaller pulley. The drivers were separately powered by 220 volts of alternating current and 4 amperes. The motor-spool combinations pulled at a maximum speed up to one and a half feet (0.5 m) per second but were operated at around 10 inches (25 cm) per second to reduce the load. The available motor torque decreases with speed, so the selection of gearing and spool size was important. The spools had to match the capability of the motors in that a small spool could pull the line with significant force, but only very slowly, while a larger spool might require the motor to

move slowly to avoid running out of torque and slipping. Some solutions to these issues could be handled from a standpoint of hardware; some could be handled with software. For instance, the torque issue mentioned above could be handled by the software telling the motors to ramp up and ramp down along their path, lessening the peak loads caused at acceleration and deceleration.

The software allows the operator to define the range of motion, maximum speed, and acceleration for each motor, and then choreograph the motors as a percentage of full range. Once each motor is initialized, the interface of a simple set of slider bars allows for easy manipulation of the variables. The software allows the track motors to turn about six hundred rotations at high speed, while the geared motors turn only fifty rotations at modest speed and the nongeared spooling motors turn only about twenty rotations. Each driver circuit board controls six motors. All eight-driver circuit boards share a single serial data channel. The command format includes a two-number address to select which circuit board and which motor on the circuit board will respond to the command. Commands may be sent to all motors at the same time with a global command. All commands expect units in ticks, except velocity and acceleration commands, which are measured in ticks per second and ticks per second squared. The number of ticks per revolution is set with switches on the power driver mounted next to the motor. Floor motors were to be set to 200 ticks per rotation, while the overhead motors were set to either 1,600 or 3,200 ticks per revolution.

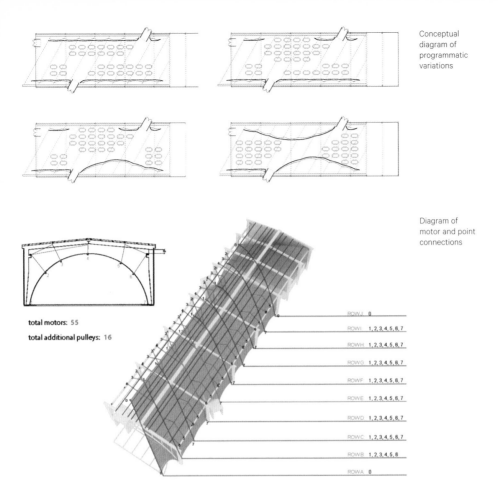

Conceptual diagram of programmatic variations

Conceptual diagram of programmatic variations

Diagram of motor and point connections

total motors: 55

total additional pulleys: 16

ROW J   0
ROW I   1, 2, 3, 4, 5, 6, 7
ROW H   1, 2, 3, 4, 5, 6, 7
ROW G   1, 2, 3, 4, 5, 6, 7
ROW F   1, 2, 3, 4, 5, 6, 7
ROW E   1, 2, 3, 4, 5, 6, 7
ROW D   1, 2, 3, 4, 5, 6, 7
ROW C   1, 2, 3, 4, 5, 6, 7
ROW B   1, 2, 3, 4, 5, 6
ROW A   0

Additional decisions included the scaling of the spools from 1 inch to 0.4 inches (2.5 cm to 1 cm) and using a braided line rather than steel cables on most of the motors; Spectra line is made of ultra-high molecular weight polyethylene (UHMWPE) fibers that yield very high strength and are cut resistant, with much greater flexibility and one-tenth the weight of steel. Although the geared motors have steel cables for safety reasons, it was found that the steel cables are much more problematic in terms of the connections and wear over time. The only major issue with the fabric was that it could be pre-stretched a bit more, to 15 percent, in order to reduce sagging between the ribs; although the poles at the base were bending, pre-stretching was an issue that could be resolved by adjusting the pulley-point locations in the ceiling. The final construction included many smaller aesthetic considerations and was coordinated with projection mapping onto the fabric, acoustic design, and general event lighting.

In general, a project such as this is much more difficult to create than static architecture, primarily due to the lack of many multidisciplinary precedents. The process will certainly get easier by virtue of the lessons learned on such built projects and the capabilities of

Physical models demonstrating transformational capabilities

Motor pair configuration

the environments they produce. The project described here begins to map out a world wealthy in its potential for motion, a world in which spaces and objects can move and transform to facilitate numerous changing situations, from the contextual and environmental to the programmatic. Our capabilities for using kinetics in architecture today can be extended far beyond what has to date been possible. Advancement, however, will only be accomplished when we address kinetic structures not primarily or singularly, but as integral components of a larger architectural system.

DYNAMIC LINES | continuous line

dotted line

SMART HIGHWAY
Studio Roosegaarde and Heijmans Infrastructure

Smart Highway is an inclusive project consisting of interactive and sustainable roads developed by artist Daan Roosegaarde and builder and developer Heijmans Infrastructure, whose collaboration is an example of true industry. The design and interactivity of Roosegaarde and the specialized knowledge and craftsmanship of Heijmans bring together the best of both worlds, fusing into one common goal: the innovation of the Dutch landscape.

A lot has been written about intelligent highways since the 1980s. Until now, however, the focus of innovation has been on the car. Studio Roosegaarde and Heijmans tackle this issue on a large scale by innovating the road deck with their application designs: Glow-in-the-Dark Lining, Dynamic Paint, Electric Priority Lane, Interactive Light, Dynamic Lines, and Wind Light; new designs include Induction Priority Lane and Road Printer. Their aim was to create smart roads by using light, energy, and road signs that interact with traffic situations. Sustainability, safety, and perception are key to the concept and manifest in the latest technologies in energy and light and in the several custom-made applications behind the smart roads.

Glowing Lines is the implementation of a Glow-in-the-Dark Lining that absorbs energy

GLOWING LINES

charging

shining

during the day and glows in the dark at night. The lining emits light for as long as ten hours. The concept is a safe and sustainable alternative to conventional lighting for dark roads. They now feature along the N329 highway in the city of Oss, the first "road of the future," and the project is planned to launch internationally.

Dynamic Paint is a similar concept of temperature-controlled marking that lights up and becomes transparent again depending on the temperature. The marking warns road users when the road deck can be slippery. Drivers experience direct interaction with the road deck.

The Electric Priority Lane is an induction charging lane that offers electric cars the option to charge while driving. Electric Priority Lanes support and stimulate sustainable transportation.

Interactive Light is interactive lighting that is controlled by sensors: it only turns on when traffic approaches. It is a sustainable and cost-saving alternative to continuous lighting. Interactive lighting can also provide speed guidance.

Dynamic Lines are road-deck markings that can be adjusted to show a continuous line or a dotted line. This is a two-part strategy whereby

traffic control adjusts to the situation and Dynamic Lines facilitate capacity management.

Wind Light is the last of the concepts to be implemented in the *Smart Highway* project. For this concept, the wind generated by passing cars activates small windmills along the road. The windmills generate energy, which is then used to light the lamps in the windmills. Wind Light is an energy-neutral application that makes the contours of the road visible, adding dimension to the interactive experience of the highway.

Smart Highway has been awarded a Dutch Design Award and an Accenture Innovation Award, and it is the winner of the INDEX Award 2013. It is an innovative concept for smart roads of tomorrow, a program of innovation that links a different way of looking at things with the opportunities afforded by new technologies.

ELECTRIC PRIORITY LANE

INTERACTIVE LIGHT

Wind Light

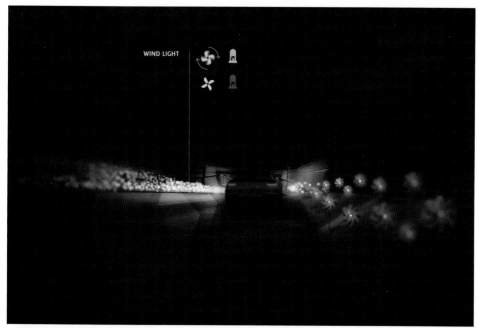

WIND LIGHT

Dynamic Paint

DYNAMIC PAINT

-5°

0°

20°

EVOLVE

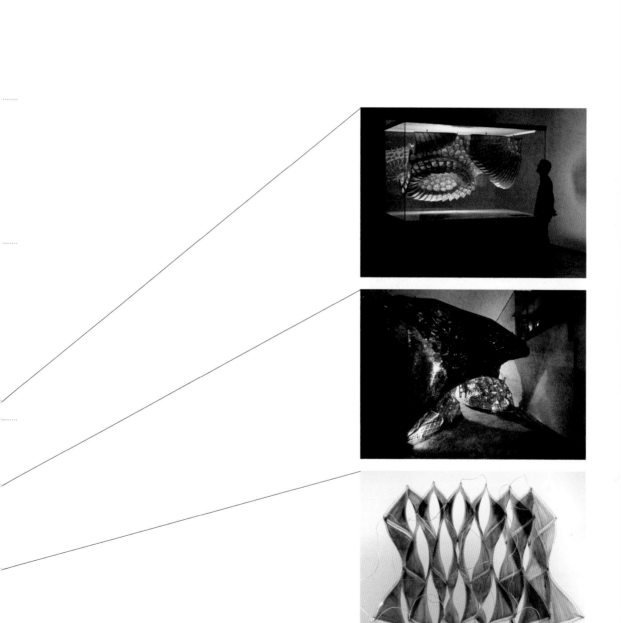

The architect and theoretician John Frazer points out that an architect's blueprint is a specific one-off set of plans, whereas nature's blueprint is a set of instructions that depend on a particular environmental context for their interpretation. Our present search to go beyond the blueprint in architecture and to formulate a coded set of responsive instructions (what we call a "genetic language of architecture") may yield a more appropriate metaphor.[1] Architectural design has been fascinated by evolutionary concepts developed in the recent past. While such explorations have lately contributed a great deal to the theory and practice of architecture, the conversation encircling evolutionary concepts has typically been constrained to the design processes and ends with the building of the building. In other words, the concepts have been used in the design of many buildings and environments, but not in the buildings and environments themselves.

The projects in this chapter are concerned with the concept of evolutionary design at precisely that moment, once the projects are being built. While the architectural world has embraced evolutionary

systems from the standpoint of computational design, the fascinating potential as an integral part of actual built form is still in its infancy. The view that evolutionary processes position built architecture as a living and evolving system relies heavily on biological and scientific analogies as well as cybernetics, complexity, and chaos theory. Frazer outlines eight aspects of evolution: development through natural selection, self-organization, metabolism, thermodynamics, morphology, morphogenetics, symmetry breaking, and the prevalence of instability. All of these aspects of evolution produce change at a variety of scales, but the basis of all such conditions is information.[2] The aim of these projects is to achieve the symbiotic behavior and metabolic balance found in nature. To do so, architecture must operate like an organism, in a direct analogy with the underlying design process of nature. The great unknown in our built environments, despite the best intentions of the designers, is the users—in particular, users that become active participants. Computer scientist and business architect Richard Veryard observes that "architecture-in-use emerges from a complex set of interactions between the efforts and intentions of many people. The architects cannot anticipate, let alone control, all of these interactions."[3]

In the recent past, we have seen a wealth of explorations in biomimetics applied to architecture. The profession has been rapidly embracing digital design technologies applied within a framework of biologically inspired processes. The physicist Yoseph Bar-Cohen sums it up well: "Nature has 'experimented' with various solutions and over billions of years it has improved the successful ones."[4] He adds: "Adapting mechanisms and capabilities from nature and using scientific approaches led to effective materials, structures, tools, mechanisms, processes, algorithms, methods, systems and many other benefits." In terms of the development of robots, he states that "the multidisciplinary issues involved include actuators, sensors, structures, functionality, control, intelligence, and autonomy."[5] Put simply, nature is the largest laboratory that has ever existed and ever will exist. In addressing its challenges through evolution, nature tested every field of science and engineering, leading to inventions that work well and last.

Technological advancements in manufacturing and fabrication, in particular with respect to materials, have continued to expand the parameters of what is possible in the area of natural adaptation. These advancements influence the scale on which we understand and construct our world, resulting in a reinterpretation of the mechanical paradigm of adaptation. Such advancements have fostered an understanding of adaptation that is more holistic and operates on a very small internal scale.

A biological paradigm requires more than just understanding pragmatic and performance-based technologies; aesthetic, conceptual, and philosophical issues relating to humans and the global environment must also be taken into consideration. A number of architects and philosophers are beginning to formulate the basis for a physically dynamic

architecture that is supported by an improved understanding of biological systems and of scale in particular. Arnim von Gleich and his colleagues at the University of Bremen have identified three main strands of development in biomimetics: 1) functional morphology (form and function); 2) signal and information processing, biocybernetics, sensor technology, and robotics; and 3) nanobiomimetics (molecular self-organization and nanotechnology).[6] The organic paradigm reinterprets the scale at which designers work and view the world at at all scales. In the recent past we have seen innovations in related fields of automotive and aerospace engineering derived from electronic systems, but now we are beginning to witness an explosion of innovation in manufacturing and fabrication that is heavily influenced by both biology and scale. Until recently, robots were getting smaller and smaller and relying on the tiniest of conventional mechanical parts. From a material vantage point, the possibilities at hand make the mechanical paradigm seem dated, ironically, before it ever had a chance to fully manifest. As the architect Michael Weinstock poetically states: "Material is no longer subservient to a form imposed upon it but is instead the very genesis of the form itself."[7]

Developments in architecture have always been intrinsically tied to developments in materials. The architect Toshiko Mori has pointed out, "we can theoretically produce materials to meet specific performative criteria; this transformation often takes place at the molecular level, where materiality is rendered invisible."[8] *Intelligent* and *smart* are general terms for materials that have one or more properties that can be altered. The architect Blaine Brownell describes transformational materials as those that undergo a physical metamorphosis when triggered by environmental stimuli; such change may either be based on the inherent properties of the material or user driven.[9] The architects Michelle Addington and Daniel Schodek divide smart materials and systems into two classes. "Type one" materials undergo changes in one or more of their properties (chemical, electrical, magnetic, mechanical, or thermal) in direct response to a change in external stimuli. A "Type two" smart material transforms energy from one form into another; this class involves materials that exhibit the following types of behavior: photovoltaic, thermoelectric, piezoelectric, photoluminescent, and electrostrictive.[10]

Although they are not common in architectural scenarios, many other industries are already demonstrating how smart materials can be used, as sensors, detectors, transducers, and actuators. As a piezoelectric material is deformed, it gives off a small but measurable electrical discharge. An example of this kind of material is the airbag sensor in your car, which senses the force of an impact on the car and sends an electric charge to deploy the airbag. The architect and professor John Fernandez points out that such embedded sensors, self-healing composites, and nanoscale and responsive materials are perfectly poised for many architectural applications, because they can counteract loads and reduce material, change shape to block sunlight, allow for active ventilation and insulation, and prevent their own degradation.

The projects featured in this section demonstrate a variety of approaches to adaptation at the level of materials. The beauty is that none of them relies on an elaborate technical system, and, as Achim Menges aptly puts it, "they show how an intellectual investment in higher design integration, allows for the realization of strikingly simple yet effective systems." Menges and the computational designer Steffen Reichert have been fascinated with the particular biological processes of hygroscopic actuation and applied them, with astonishing results, to a couple of their projects. Based on the relative humidity, a material change occurs; this is a decidedly no-tech approach in which the system's adaptive capacities are ingrained in the material. *HygroScope: Meteorosensitive Morphology* is self-contained within a regulated environment, and *HygroSkin: Meteorosensitive Pavilion* is in an unregulated outdoor environment. Doris Sung has taken a very similar approach with *Bloom* but uses thermobimetals to create a dynamically responsive skin with the goal of self-regulating the building temperature. Bimetals are formed by laminating two thin sheets of metals that expand at different temperatures, causing the sheets to curl when heated. Sung's project also makes an important contribution in scalar differentiation. She puts it as follows: "A brick wall is the same shape over and over again, but if you look at a fish, each scale is a unique size and conforms to its specific location." Also significant is that the project has been applied at an architectural scale and proven itself over time in the fluctuating outdoor environment.

Manuel Kretzer's *ShapeShift* demonstrates a number of interesting explorations with electroactive polymers, soft plastics that change their size, shape, or volume in response to a strong electrical field. Despite decades of research and continual progress, the domain of electroactive polymers is still far from mature, and a number of topics, such as long-term performance and durability, need immense further development. Nevertheless, Kretzer has demonstrated a number of potential applications that are based on the shape flexibility of electroactive polymers. In other professions, says Kretzer, these polymers "have already been utilized as artificial muscles, both for prosthetic purposes or 'super-human' exoskeletons, microelectromechanical systems (MEMS), medical devices, speakers and more concrete applications, such as refreshable braille displays, or gaming interfaces." Although there is still work to be done regarding the feasibility of architectural durability, there is great potential for innovation. An awareness of advancements in both robotics and new materials promises an architectural future in which adaptation is more holistic and operates on a small internal scale.

HYGROSCOPE + HYGROSKIN
Achim Menges
with Steffen Reichert and Oliver David Kreig

The *HygroScope* installation as a permanent exhibit at the Centre Pompidou in Paris

Biology offers an unmatched level of adaptation and responsiveness from which to draw inspiration. For architecture, botany exists as a particularly interesting field of investigation, as it offers a rich repertoire of adaptive movements made without muscles. The responsive movements of plants can be categorized into two main groups: (1) active cell pressure–actuated systems, and (2) passive systems that are fully independent from a plant's metabolic trigger mechanism. The conifer cone, or more specifically its cone scales, is an interesting example of the latter. In dry conditions, the cone's materiality allows it to adjust its moisture content to the environment. The related dimensional change of the material triggers a shape change in the cone scales, which leads to the release of its seeds. While many passive plant movements are irreversible, conifer cones retain their responsiveness even after the biological function of seed release is fulfilled. The material responsiveness to environmental conditions is based on its hygroscopic behavior—its ability to absorb moisture from the air and release it back, in pursuit of a constant equilibrium between its own moisture level and that of the surrounding environment. During these processes of absorption and desorption, a physical change occurs. Water

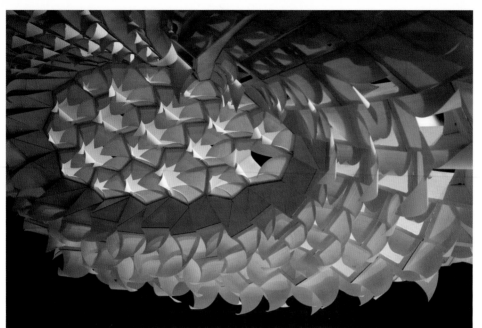

left and below:
Physical changes
reveal the dynamic
nature of the
HygroScope
project

molecules become bonded to or released from the material. The changing distance between the microfibrils within the wood cell tissue causes shrinkage and expansion. The scale movement is an effect of the combination of the material's dimensional change and anisotropic composition. Such directional characteristics can be found in the materiality of the scales' cross-layered fiber. Since all the actuation energy for the shape change is derived from the surrounding environment, conifer cones can be described as passive, autonomous responsive systems.

Since responsiveness in architecture is usually perceived as a technical function, most solutions require separate (electro-) mechanical sensing, actuation, and regulation mechanisms. The projects presented here follow a biological, no-tech approach, in contrast to high-tech mechanical endeavors. The adaptive capacity of their systems is ingrained in the material; no additional technical equipment is required—for sensing, actuating, or regulating.

Based on the biological example of the conifer cone, Achim Menges, Steffen Reichert, and Oliver David Kreig have developed a hygroscopic material system. The responsive materiality of their system is achieved with wood—more precisely, simple, quarter-cut veneer—in combination with a synthetic fiber-reinforced polymer. Wood is one of the oldest and most common construction materials, and its hygroscopic capacity is well understood yet finds almost no application. In fact, it is usually perceived as a problematic characteristic of the material. Thus, most energy in timber processing is spent on reducing the material's

moisture content and its innate tendency to move anisotropically.

The *HygroScope* project takes a different route, instrumentalizing the inherent dimensional change of wood as the trigger mechanism for an adaptive shape change. The actual movement can be controlled by varying the following material-specific parameters: (1) the fiber directionality; (2) the layout of the natural and synthetic composite; (3) the length-width-thickness ratio; (4) the element geometry; and, especially, (5) the humidity control during the production process. In this way, movement behavior can be precisely controlled and calibrated to specific ranges of relative humidity change. The response behavior can even be manipulated in such a way that the same basic material element can be tuned to either open or close when exposed to the same stimulus. The decomposition of the cone's complex structure and the transfer into a set of material and fabrication parameters and rules formed the basis of the computational design strategies of *HygroScope*, as well as the related ability of physically programming the material behavior in response to the specific project context.

HygroScope: Meteorosensitive Morphology was commissioned by the Centre Pompidou in Paris for the museum's permanent collection and was first shown in the exhibition *Multiversités Créatives* in May 2012. The installation displays an autonomous, responsive architectural surface suspended within a climate-controlled, fully transparent glass case. The visually floating surface follows the no-tech paradigm of the related research and opens and closes in response to its surrounding humidity. Located in possibly one of

The hygroscopic material system is based on the biological example of the conifer cone the most stable, climate-controlled spaces in the world—the interior of the Centre Pompidou—the glass case houses a custom climate-control unit that simulates the humidity changes of the city outside the building. The related oscillation of moisture level causes the system to silently open and close, enabling the visitor to experience the fluctuation of relative humidity that forms an intrinsic part of everyday life but usually escapes our conscious perception. At the same time, the installation triggers an awareness of untapped energy sources latent in the dynamics of our environment.

The morphological articulation of the system was developed through a custom computational process that incorporated the reciprocal system-intrinsic parameters, the fabrication constraints of a state-of-the-art robotic fabrication technology, and extrinsic environmental considerations. This integrative design methodology allowed for exploring the morphospace of possible design variations. The installation was constructed from more than four thousand geometrically differentiated elements and employs wood as the primary construction material. The global shape shows three protuberances of continuing surface, which are connected by a cellular channel area. The different areas are designed to form separate microclimate zones that correspond to the humidity-release locations. The resulting two morphologically distinct zones of the system are equipped with two different opening mechanisms: (1) the center of the protuberances, which act as the channel's centralized opening mechanism, and (2) the second opening mechanism, whereby the radially ripped main bodies of the protuberances

form a transition between the center areas and the channels and are covered with a directional configuration that opens toward the channel areas. In low-humidity conditions, the structure forms a fully closed surface, but with an increase of moist air, the surface transforms into a porous system. The related response requires neither electric nor mechanical equipment nor a supply of energy. The veneer-composite elements are at once sensor, motor, and surface.

HygroSkin: Meteorosensitive Pavilion was commissioned by the Fonds régional d'art contemporain présente ses collections et expositions (FRAC) Centre Orléans for its permanent collection and was first shown as part of ArchiLab's Naturalizing Architecture exhibition in 2013. The project explores the tension between an archetypical architectural volume, the box, and a deep, undulating skin that embeds clusters of intricate climate-responsive apertures. The traveling pavilion's modular design is based on the stringent requirement of a small packing volume. The modules themselves are derived from the elastic bending behavior of thin plywood material and the potential for generating global structures from a common conical base geometry.

Plywood can be elastically bent into developable shapes such as cylinders or cones, each conical element consisting of a double-layered skin. First, the two separate cones are elastically self-formed. Next, a sandwich element is produced using vacuum pressing, laminating the plywood pieces onto a foam core. The initial conical modules are surveyed for tolerances that use a seven-axis industrial robot. The modules' edges are trimmed into their final differentiated form

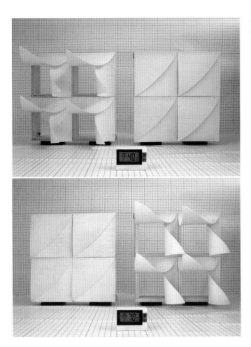

Component models
of the hygroscopic
material system
in different
responsive states

using the robot with a circular saw tool. Finally, the inner foam core is milled into shape, resulting in lightweight, differentiated modules that can be assembled into a robust, structural system yet easily transported. Within the concave surface of the conical modules, injection-molded aperture elements are placed, forming the support structure for the weather-responsive material. In contrast to the previous *HygroScope* project, the responsiveness of the apertures was physically programmed to close with an increasing humidity level and serve as autonomous weather protection. Calibrated to a range of 30 to 90 percent relative humidity, the apertures utilize the full spectrum of moderate climate zones. In direct feedback with their local environment, they continuously adjust their degree of openness and porosity, modulating the light transmission and visual permeability of the envelope. The pavilion is therefore designed as a unique convergence of environmental and spatial experience.

Together the projects presented here form a first step toward responsive architecture that does not rely on elaborate technical systems. In many ways, they show how an intellectual investment in higher design integration allows

for the realization of strikingly simple yet effective systems. In addition, they suggest a promising coalescence of system performance and architectural performativity, as the complex environmental dynamics lead to continuously differentiated spatial conditions. While the *HygroSkin* pavilion is programmed to close with increasing humidity and serve as a meteorosensitive weather shelter, the *HygroScope* project opens in response to the same stimulus, breathing and releasing humidity through the increased porosity. In controlled laboratory conditions, a linear dependency between the degree of relative humidity and the degree of openness can be observed. But being exposed to much more complex environments, both projects show fascinating unexpected local differentiation in responsiveness. Since relative humidity is defined as a function of absolute humidity and temperature, heat radiation or other ambient climate influences are responsible for the sometimes counterintuitive response of the hygroscopically actuated systems. Their behavior remains unpredictable, therefore, and enables a visual experience of otherwise hidden environmental patterns.

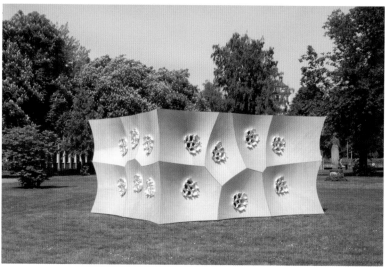

this and the following spread: Various views of the completed *HygroSkin: Meteorosensitive Pavilion*

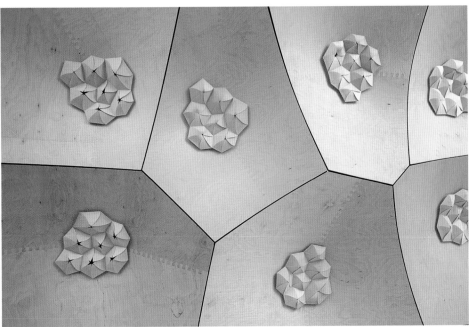

BLOOM
DOSU

A sun-tracking instrument indexing time and temperature, *Bloom* stitched together material experimentation, structural innovation, and computational form and pattern making into an environmentally responsive installation. The responsive surface was primarily made out of more than nine thousand smart thermobimetal tiles, no two pieces of which are alike. As a lamination of two metal alloys with different coefficients of expansion, a thermobimetal automatically curls when heated by the sun or by ambient temperature changes; the result was a highly differentiated skin system that could smartly shade or ventilate specific areas under the canopy without

additional power or controls. In order to heighten the sensitivity of the skin, the overall form was oriented toward the sun's arc to maximize solar exposure, much as growing plants and flowers are. The effect was further optimized by the use of powerful software to generate parametric patterns. With today's digital technology and driving interest in sustainable design, thermobimetal, a simple material, transcends its limited role as a mechanical actuating device to become a dynamic building-surface material while expanding the discourse of performative architecture.

In addition to the cultural and intellectual implications, several inherent constraints of the

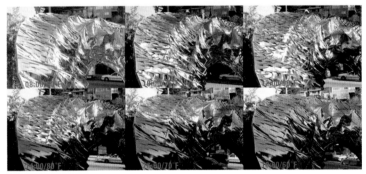

Views of the installation showing *Bloom*'s human scale and natural kinetic actuation

physical context contributed to the development of the project. Shaded by the adjacent buildings, the small outdoor courtyard at the Materials & Application Gallery in Los Angeles presented an unfortunate challenge for a project that relies on heat and sunlight to operate. Like a plant, the form grew out of the shadows and toward the sunlight, but also needed to resist gravity and lateral loads. Without the ability to attach to an adjacent building or build extensive foundation systems, Doris Sung's firm, DOSU, put extensive effort into finding a form that would be light-weight, structurally sound, beautiful, and performative. The product was a canopy that blooms

toward the sun, provides a dynamic urban area below, and assumes a shape representative of our technological age and renewed interest in biomimetics. It references nature in its form, in its movement, and in its name.

The main goal of *Bloom* was to demonstrate the efficacy of thermobimetals as an exterior building surface with two functions. The first of these involved the bimetals' potential as a sun-shading device that dynamically increases the amount of shade with the rise of the outdoor temperature. The size, shape, and orientation of the tiles were positioned strategically to perform optimally according to the angle of the

The pavilion in
its dense urban
context

sun by use of advanced modeling software. The second function of the bimetals was to ventilate unwanted hot air. Optimizing the contortion of individual bimetal tiles caused any captured heat to trigger the surface tiles to curl and passively ventilate the space below. The use of complex digital tools such as Computer Aided Three-Dimensional Interactive Application (CATIA), Rhinoscript, and Grasshopper with other solar and structural analysis tools continues to challenge the traditional processes of design and the capabilities of fabrication. With surprisingly few mistakes in the final fabrication files, the use of digital software in this proof-of-concept installation proved to be an amazingly useful tool in computer-aided manufacturing.

Composed of 414 hyperbolic paraboloid–shaped stacked panels, the self-supporting structure tested the capability of the materials to perform as a shell in a completely unique way. The panels combined a double-ruled surface of bimetal tiles with an interlocking, folded aluminum-frame system. In some areas of *Bloom*, the hypar panels were stiffened by increasing the number of riveted connections, while in other areas, the peaks and valleys were made deeper

(and more twisted) to increase structural capability. The final lightweight and flexible monocoque form was dependent on the overall geometry and combination of materials to provide comprehensive stability. Once complete, the structure automatically conformed to its designed geometry and gains maximum stability. This method of design allowed the project to weigh in at an estimated six hundred pounds—lightweight in the structural-shell world of design.

With the availability of digital-to-digital fabrication, most of the design-build effort was invested in the preparation of the digital models and fabrication files. All construction joints were reduced to simple rivets or nut/bolt connections so that special tools or costly methods of construction are eliminated. The repetitive connections were assembled using wrenches, pliers, and rivet guns—tools anyone can use. Additionally, the need for large-scale equipment was also eliminated by strategically building the structure from the top down, allowing all hand assembly to take place on the ground. To do this, a large wooden ring—slightly larger than the topmost edge of *Bloom*—was needed as an armature to hang the overall piece during

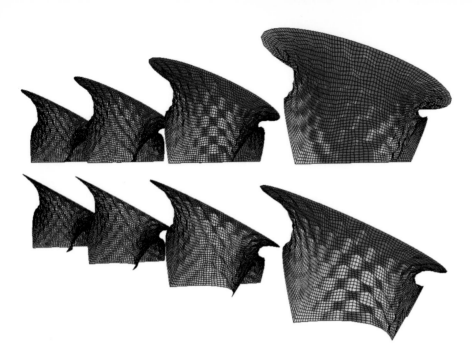

left and opposite:
Architectural
drawings showing
the indexing
of time and
temperature.

construction. Row by row, panel by panel, the project slowly took shape. The ring was raised higher and new rows of panels were inserted from below until the final row locked in the overall geometry. Until the structure was settled on the ground, the surface could not take its true geometry and full strength.

The impact of *Bloom* on the design profession, construction industry, academia, and general public should be a paradigm shift: the project asserted a sustainable, passive method of reducing reliance on artificial climate-control systems and, ultimately, on the waste of valuable energy. It exhibited an innovative structural strategy that valued distributed structural stresses and reduced infrastructural needs, and it demonstrated the power of digital technologies in the design, analysis, and fabrication of complex tessellated surfaces. Currently, new alloy laminations are being developed in collaboration with manufacturers for durability and for greater deformation characteristics in shape-memory alloys, while other architectural applications, like building screens and *brise-soleils*, are beginning production. Because *brise-soleils* have the potential for immediate application, patterning

and tile shape play a large role in performance and production. Different from their custom-made counterparts, mass-produced building components benefit from the insertion of bimetals in their composite systems. In double-glazed panel systems, the bimetals can behave like an automatic shutter system requiring no energy, while in cement blocks they can allow air to pass through when temperatures are high to make breezing walls. In all cases, the tangible contribution to sustainable design is clear, and the variety of continued research projects is a testament to the potential of this smart material in performative building skins.

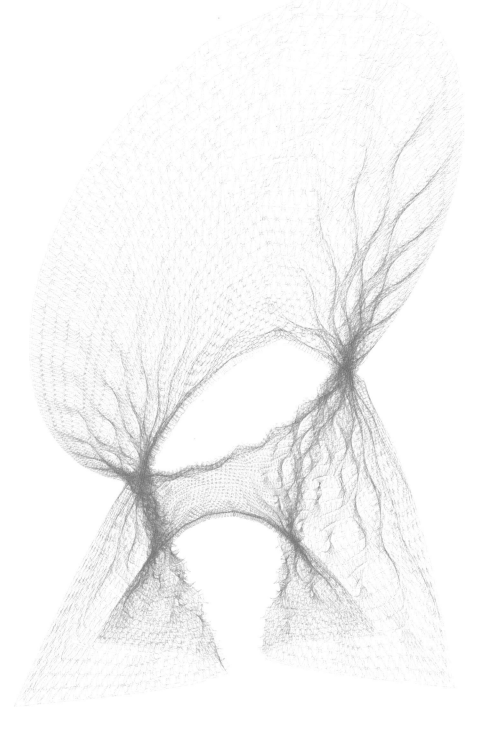

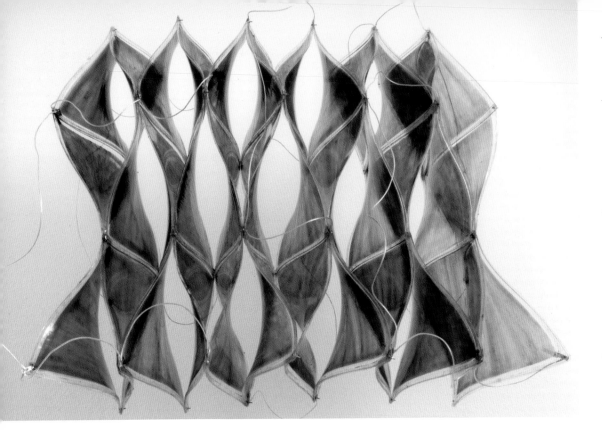

SHAPESHIFT
Manuel Kretzer

ShapeShift investigated the potential of electroactive polymers (EAP) to create a dynamic spatial intervention. Electroactive polymers are soft plastics that can change their size, shape, or volume in response to a strong electrical field. In the domain of active materials, they stand out due to their large deformation potential, high response speed, low weight, and price. While EAP research mainly focuses on their application as artificial muscles, *ShapeShift* wanted to highlight their quality as dynamic, space-forming surface material.

According to their working principle, EAPs can be split into two main categories: ionic and electronic electroactive polymers. Ionic EAPs are driven by a displacement of ions during electrical stimulation, which leads to a change in shape or volume. Their main advantage is that they can be actuated by voltages as low as one to two volts; however, since the ions are diffused inside an electrolyte, they need to maintain their wetness at all times. For their strong bending capabilities, they are mostly used as bending actuators, where high forces are required but have a rather slow response speed. Since the production of stable material configurations requires high precision, they are expensive and usually not commercially available.

Electronic EAPs are driven by strong electric fields. The electrostatic forces lead to an electro-mechanical change in the shape of the material. Usually, electronic EAPs are applied as planar actuators due to their large in-plane deformations. In contrast to their ionic counterparts, they work in dry conditions but need very high activation voltages, in the range of several kilovolts. They exhibit a very short response delay, display a relatively large activation stress, and can hold the induced displacement under direct current (DC) activation.

The first reported occurrence of an electroactive phenomenon dates back to 1880, when Wilhelm Röntgen observed a length change in a rubber band, fixed at one end with a weight that was electrically charged and discharged. In 1899, M. P. Sacerdote confirmed Röntgen's experiment and formulated a theory of the strain response to electric-field activation. In 1925, electret, the first piezoelectric polymer, was discovered by combining carnauba wax, rosin, and beeswax, then, while cooling and solidifying the mixture, exposing it to a DC bias field. In 1969, it was demonstrated that polyvinylidene fluoride (PVDF) displays a large piezoelectric constant. In 1977, Hideaki Shirakawa

doped polyacetylene with iodine and thereby enhanced its conductivity by eight times, getting it close to that of metal. In the early 1990s, ionic polymer-metal composites were developed, and they displayed a much lower activation voltage while sustaining a larger deformation than previous electroactive polymers.

The first commercially developed device containing EAPs as artificial muscles was a fish, produced by Eamex in Japan in 2002. It worked without batteries or a motor, relying simply on EAP materials that bent when energized by inductive coils at the top and bottom of the fish tank. In 1990, the Defense Advanced Research Projects Agency (DARPA) funded research that led to the development of an EAP based on silicone and acrylic polymers. The company Artificial Muscle further developed the technology in 2005 and started industrial production in 2008. In 2010, Artificial Muscle became a subsidiary of Bayer Material Science. Since 2008, Danfoss manufactures and commercializes a dielectric EAP material called PolyPower, which is mostly intended for use as variable stretch sensors.

Among the electronic EAPs, the especially soft dielectric variety (also referred to as

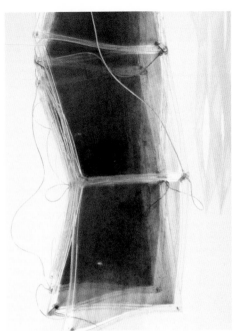 

dielectric elastomers) exhibit promising functional properties. They are capable of strains up to 380 percent, are extremely flexible, light, thin, and transparent, and can basically be tailored to any size or shape. The central element of dielectric elastomer actuators consists of a thin elastomeric film, made of silicone or acrylic. The membrane is coated on both sides with or is sandwiched between two compliant electrodes. In this configuration, the polymer acts as a dielectric in a compliant capacitor. When an electrical voltage is applied, opposite charges move from one electrode to the other and squeeze the film in its thickness direction, which leads to a

planar film expansion. Once the voltage is turned off and the electrodes are short-circuited, the EAP returns to its original shape.

Two main types of dielectric elastomers have so far been developed. The stacked, or contractile, actuators consist of several hundred or thousand layers and can work against external tensile loads, acting in the thick direction since the material is compressed when activated. The expanding actuators perform work against external pressure loads in both planar directions. They are usually under a certain prestretch and are often embraced by a mechanical support structure or frame. In these systems, the

Activated expanding actuator component

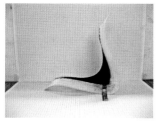 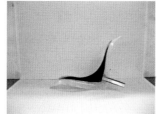 

Iterative studies of size, shape, stretch, frame, and material relationships

support is released when the material relaxes under activation.

In order to successfully create a dielectric elastomer actuator, particular requirements have to be met. The film must be thin with uniform thickness, incompressible, and a good electrical insulator. The frame must be flexible yet strong enough to maintain the pre-stretch and pull the film when actuated. Most of all, electrodes need good conductivity, even when stretched; they must be as thin as possible and bond well with the dielectric film. This is usually achieved by spraying or brushing electrically conductive particles, such as metal, carbon black, or graphite powder, onto the membrane. In order to supply the dielectric elastomer with the necessary high-DC activation voltage, high voltage sources or amplifiers are utilized. For reasons of longevity and safety, the finished component can be insulated with a passive silicone layer, though it reduces the component's active capacity.

ShapeShift emphasized the aesthetic properties and the soft organic movement of the material through an interactive spatial installation. The structure consisted of thirty-six identical active elements, produced as described above.

Once a DC voltage in the range of three to five kilovolts was applied, the film was compressed in its thick direction (against the vertically stacked layers), which led to a planar expansion of the membrane. Since the membrane was attached to the flexible frame, the frame bent when the material was in its relaxed state and flattened out when the tension was removed during actuation.

Through numerous design iterations, the students altered the shape, size, frame-border width, material thickness, and materiality of the frames until the movement was empirically maximized and the desired three-dimensional motion was achieved. Parallel to the development of single components, investigations into structural arrangements were performed. Due to the laborious manual fabrication procedure, tessellations based on a minimal number of elements were explored. In the end, a very simple arrangement, derived from a multitude of identical components, was selected. Through physically connecting the components together, dynamic configurations could be achieved that enhanced the movement of the individual elements and resulted in feasible, self-supporting structures. As with the single shapes, the final form of these

tessellations resulted from the relationship of the dielectric elastomer to its frame and the connections to neighboring elements.

A number of potential applications have emerged, based on the shape flexibility of EAPs, which, among other qualities, makes them highly versatile materials. They have been used as artificial muscles, both for prosthetic purposes and superhuman exoskeletons, microelectro-mechanical systems (MEMS), medical devices, speakers, and, more concretely, as refreshable braille displays or gaming interfaces. Based on their immediacy and natural or organic behavior, recent research has investigated their potential for active camouflaging on soft surfaces, mimicking the principles of the chromatophores of cephalopods (squid, cuttlefish, octopus). Chromatophores are color-changing cells found in the skin of the animal. Even though a large variety of EAP actuators has been successfully demonstrated, most materials are still custom-made, and the number of commercially available products is limited.

In addition to acting as mechanical actuators, EAPs can perform as sensors. When the charged EAP device is stretched or contracted, the serial resistance within the electrodes changes. Furthermore, they can be used to convert mechanical work into electrical energy and, as such, be applied as dielectric generators—for example, to harvest energy from sea waves—with their softness and flexibility in shape and design providing additional qualities. More likely, however, will be the use of EAPs for portable devices and peripherals. They predict a haptic usage in the consumer electronics touch-display market exceeding 60 percent until 2018, and that will account for more than 40 percent of the expected total revenue in five years.

For architectural applications, EAPs exhibit very promising properties; their homogeneous surface quality, transparency, and large, active deformation are visually astonishing. Problematic, however, is their scalability and especially their durability, which cannot yet comply with architectural standards, especially when exposed to changing environmental and climatic conditions. Automating the manufacturing process could partially help to make more consistent components, but in order to make them more stable, a carrier material other than the Very High Bond (VHB) film by 3M would

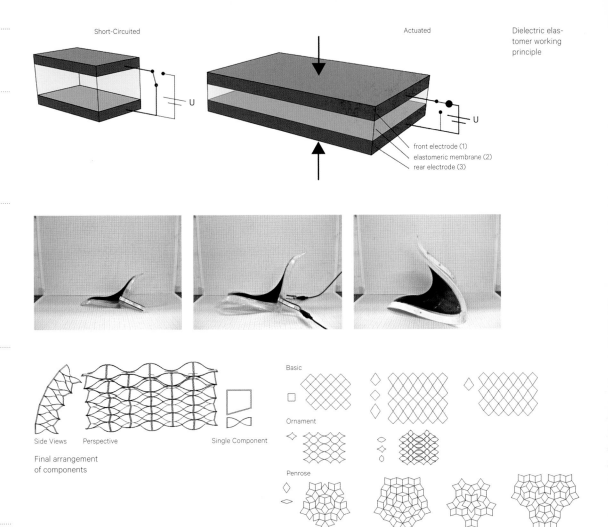

Short-Circuited

Actuated

Dielectric elastomer working principle

front electrode (1)
elastomeric membrane (2)
rear electrode (3)

Side Views   Perspective   Single Component

Final arrangement
of components

Basic

Ornament

Penrose

Various tiling
options during
development

have to be developed. In public spaces also, the
accessibility, location, and possible insulation of
the material need to be considered. The components require fairly high voltages, which pose a
certain danger and can harm people when they
touch the elements during activation.

CATALYZE

[
Definition: to bring about, inspire
Related Words: conduce (to), contribute (to); begin, establish, father, found, inaugurate, initiate, innovate, institute, introduce, launch, pioneer, set, set up, start; advance, cultivate, develop, encourage, foster, nourish, nurture, promote

Merriam-Webster OnLine, s.v. "catalyze," accessed March 25, 2015,
http://www.merriam-webster.com/dictionary/catalyze.
]

Philosophically, interactive architecture is in a unique position to reposition the role of the designer. This role should be less about creating a finished design and more about catalyzing—about asking how a design may evolve. In a sense, designing interactive architecture should be an egoless, emergent endeavor that lies in designing the platform for the future, not the future itself. Such a position is both noble and profound, for it means the designer must understand people well enough to be able to design for them yet also design interfaces and tools such that people can in turn become designers. What has made the ubiquitous smartphone so powerful is not that it is a connected device, but that it is a platform for the creation of applications. It has become a catalyst for design and ideas that were never intended.

To preface, let's begin with a bit of clarification in terminology. As the term *adaptive* has increasingly fostered a shift from the paradigm of the mechanical to the biological, it has left a gap in the area of control. In the strictly biomimetic approach to architectural design, robotics helps to fill this gap. Biorobotic architecture defines a biomimetic adaptation that is augmented with robotic control. The projects in this area define an architecture that goes beyond the mere capacity to interact to reposition designers as catalysts who can adapt and evolve with the world around them. A cybernetic system is an inclusive one, encompassing organism, machine, organization, and environment. As mentioned in the introduction, Norbert Wiener characterized cybernetics back in 1948 as the scientific study of "control and communication in the animal and the machine."[1] As a secondary matter of clarification, biomimetics was created by Otto Schmitt in 1969 as a scientific approach to systems, processes, and models in nature, imitating them in order to solve human problems. *Biocybernetics*, then, is a relatively new term for an abstract science centered on the application of cybernetics to biological science. Last, it seems necessary to define *robotics*, which is here interpreted as a mechanical agent guided by a program. The point of clarifying such terminology is to focus this exploration on how robotics can serve as a means of augmented control for natural biomimetic adaptation.

Early use of pick-and-place robotics primarily utilized a system of top-down control when applied to architectural scenarios. Modular robotics, on the other hand, relies on decentralized control. Although there may be no centralized control structure dictating how individual parts of a system should behave, local interactions between individual modules can lead to the emergence of global behavior. Most architectural applications are neither self-organizing nor equipped with higher-level intelligence functions of heuristic and symbolic decision-making abilities. Most applications do, however, exhibit a behavior based on low-level intelligence functions of automatic response and communication. The beauty of such distributed control, which we have seen in many projects in this book, is that when it is applied to a large system, there is potential for emergent behavior. An emergent behavior can occur when a number of simple systems operate in an environment that forms more complex behaviors as a collective. For instance, *Epiphyte Chamber*, one of many responsive and distributed architectural environments Philip Beesley has constructed, is composed of flexible lightweight structures that integrate kinetic functions with a microprocessing, sensor and actuator systems. In the poetic words of Beesley, "An epiphyte is a kind of plant that can grow without soil, suggesting a hovering unrooted world."

In all of the projects in this section, the environments are articulated as an interface. In terms of a resolution with respect to the sensing of human factors, recent developments in the area of interface design will eventually play a major role in how we envision our interactions with architectural spaces and objects. Interface design is heavily tied to sensor innovation and manufacturing, which has heralded the availability of previously unimaginable means for gathering data and information and for intangible forms of interaction such as gesture

and even brainwave recognition. It may soon be commonplace to embed architecture with interfaces that allow users to interact with their environments. As architectural scenarios move beyond direct sensing and response conditions, we are confronted by the need for more natural and direct means of control. Many in the architectural profession have begun to study and learn from interactive media precedents and usurp the technologies employed for controlling interactive digital environments. Technologies that allow users new means to control and interact with digital information can be broken down into three general categories: touch and multitouch, gesture, and cognitive control. To date, there are many touch and multitouch interfaces; gesture interfaces are still in their infancy with regard to architectural applications; and direct cognitive controls reside on the developmental horizon but show fascinating promise.

Gestural language is perhaps the most intriguing means of control in that it enables real physical interactions. Advancements in multitouch hardware technology are significant to architecture, because, in many cases, the gestures used to control an interface replicate closely the gestures that would be used to perform these activities in real space with tangible objects. Gestural physical manipulation of physical building components and of physical space itself is arguably the most suitable form of control relative to devices, speech, or cognition. From the standpoint of control, the gestural projects developed by myself and Allyn Polancic serve as something of an intervention on the current profusion of exploration in robotic and interactive architecture. As architectural scenarios move beyond direct sensing and response conditions, we are confronted by the need for an even more direct yet still natural means of control. Currently interactive control is all too often either handheld mobile devices or very limited sensing capabilities; there is a typical disconnect in the employment of devices to control (what should be) an interface. *Conventions of Control* demonstrates that human interaction with architecture should follow the intuitive nature of its gestural vernacular, which is device-free and allows for a variety of input streams. The project establishes architecture as an interface that allows users to directly interact with the environments via an adaptive gestural language. Interaction, or play, with the physical world is crucial to the way humans learn to socialize and understand reality. The evolution of gesture-based vocabularies to control devices, interfaces, and, eventually, the entire built environment should reflect this relationship and ideology.

Behnaz Farahi Bouzanjani takes adaptive gestural control a step further than *Conventions of Control* by abstracting the architectural environment into one that has no precedent of interaction. With *Alloplastic Architecture*, the intention was to create a space that, like any soft adaptive system in nature, could physically reconfigure itself based on user movements.. The main focus of the project is the relationship between materials, form, and interactive systems of control. As the architectural theorist Sanford Kwinter observes, "soft systems evolve by internal regulating mechanisms, yet always in collaboration with forces and effects (information) arriving from an outside source."[2] Bouzanjani has created a space that can build up an understanding of its users through their bodily gestures, visual expressions, and rituals of behavior, and respond accordingly. The project documents the crucial questions behind the design of an interactive, gutturally controlled, completely malleable structure and analyzes the decisions made throughout the design process. One of the main contributions of this project is it demonstrates that the design of a physical environment can change its shape to accommodate various performances in the space based exclusively on the user's body motion. The project presents a

working knowledge of how Microsoft Kinect and other remote sensing devices might be used more universally in order to enrich the way we interact with our environment. Bouzanjani demonstrates that the designer no longer designs the final form but rather creates an initial state with autonomous dynamic potential. Perhaps the most interesting outcome of this project is that it leads one to question the conservatism and skepticism toward technology implicit in the thinking of, for example, Martin Heidegger, whose criticism of the potentially alienating effect of technology has engendered a negative attitude toward technology in general and toward computation in particular—an attitude still prevalent in certain architectural circles.[3] *Alloplastic Architecture* shows that, far from being a source of alienation, technology may itself actually combat that distinctly modern condition.[4]

The important thing is that systems that are controlled yet control themselves on another level have the potential to catalyze design. As Bouzanjani puts it: "The architect no longer designs the final form but rather creates an initial state, introduces a set of controlled constraints, and then allows the structure to be activated to find its form in real time. What results is the emergence of unexpected shapes." Rather than literally interpreting and responding to human and environmental desires, the architectural system is allowed to take a bottom-up role in configuring itself in a malleable way. As Gordon Pask states in his foreword to *An Evolutionary Architecture*: "The role of the architect here, I think, is not so much to design a building or city as to catalyze them: to act that they may evolve."[5]

EPIPHYTE CHAMBER
Philip Beesley

Epiphyte Chamber
installed at the
Museum of Modern
and Contemporary
Art, Seoul

Epiphyte Chamber is a work that featured halolike masses of delicate mechanical veils lining a series of intimate, interconnected spaces. The resulting environment acted as a primitive social architecture, offering viewers intimate sculptural places that support small clusters of activity while interlinking larger gathering areas. A hovering filter environment composed of hundreds of thousands of individual laser-cut acrylic, Mylar, glass, and aluminum elements created diffusive boundaries between occupants and the surrounding milieu. Sensors embedded throughout the environment triggered motions that rippled out from hives of kinetic parts in peristaltic waves. These structures were organized as deeply reticulated skins that sought to maximize interchange with the atmosphere and other occupants.

Epiphyte Chamber contained the first generations of an invented structural system made from thermoformed acrylic diagrid spars that created a massive bubbling, porous infrastructure full of interlinking voids, spatially akin to sinus cavities or the tunnels in termite mounds. Hybrid soil full of turbulent qualities was interwoven throughout the spaces made by this scaffold. The soil was given a seething quality riddled with chemical exchanges, folded and fissured physical detailing, and thousands of active mechanical components

intertwined together. Frond clusters fitted with mechanisms of shape-memory alloy reacted to viewers as they approached, flexing and setting off bursts of light that stimulate the protocells and trigger chains of motion. Surrounding these clusters was a dense cloud of vessels that carried salt and sugar solutions, alternately pulling in and emitting moisture. Chains of vessels contained carbon-capture formulas that absorbed carbon dioxide–bearing air. Reactive LED lighting, using shift-register microprocessor controls, created softly rolling clouds of delicate light within the hovering filters to respond to viewers' movements. Flasks containing synthetic cells were nestled at the centers of the grouped fronds. Scent-emitting glands attracted viewers and encouraged interaction with the system as a stimulus to increased air circulation and protocell formation.

Viewers walked into an intimate space of interwoven structures and fragile canopies. As they moved through the environment, they came into contact with a dense wall of tentacles fitted with sensors. These sensors were linked to shape-memory motors mounted in the sculptural scaffolds that drove the expressive movements of the flexible structures. Reacting to movement,

tonguelike lashes reached out to passing viewers in gentle, undulating caresses. Networks of simple computational devices and sensors allowed viewers to be tracked, offering small increments of gentle muscular movement to register their presence, rippling back to the viewers and initiating the sense of a breathing, ambient architecture. Hundreds of small mechanisms, which functioned similarly to glands, pores, and hair follicles in the skin of an organism, permeated the sculpture. Sensing whiskers responded to the reach of viewers with curling and twitching responses, propelling humidified air, perfume, and organic material over fields of glands and traps.

Adjacent elements communicated with one another, spreading signals in waves. Nested loops of microprocessors networked by serial communication connections worked in layers of response that move outward from individual events and overlapped in rebounds chained together in many dozens of echoing reflexes. Parallel to the software-driven communication was a physical communication system. Lashes extending from mechanism-driven shape-memory alloy actuators occasionally brushed against proximity sensors, creating self-triggered signaling and motion that

Viewers walk into an intimate space of interwoven structures and delicate canopies

propagated in turbulent cycles. When occupant activity heightened, the structures of the space would become saturated with instability, combining both physical triggers and behavior caused by software-based communication.

Alongside the mechanized component systems, a wet system that supported simple chemical exchanges in the same way that the human lymphatic system's renewing functions operate was introduced into the environment. Thousands of primitive glands containing synthetic digestive liquids and salts were clustered throughout the system. The adaptive chemistries within captured traces of carbon from the vaporous surroundings and built delicate structural scaffolds. Engineered protocells—liquid-supported artificial cells that share some of the characteristics of natural living cells—were arranged in a series of embedded incubator flasks. Bursts of light and vibration, triggered by viewer movements, influenced the growth of the protocells, catalyzing the formation of vesicles. Organic batteries made of glass flasks containing vinegar with copper and aluminum electrodes produced tiny amounts of internally generated electricity, emitted as triggers for acoustic generators that

produced subtle, drifting veils of murmurs and whispers. Responding to trembling waves that moved throughout the layered mechanisms of the structures, viewers could become aware of subtle impacts: air moving around the body, quiet rustling, crystalline formations within suspended liquid flasks in the process of condensing and precipitating.

Epiphyte Chamber expressed a very primitive hunger for a safe home—a fundament. The kind of spaces that epiphytes create can seem emotionally charged: on one hand communicating optimism, a sense of genesis, and a place of beginning; on the other, speaking of the loss as the natural world dissolves away. Oscillating between these emotional states, this space might behave as a primordial home that wraps around the occupant and offers the gentlest kind of shelter from an atmosphere stressed by radical dispersal.

The work might evoke questions about the rather promiscuous exchanges that occur between the natural and built environments. There are personal dimensions at work here—the hovering qualities of the work echo personal bodily sensations, such as those found in the

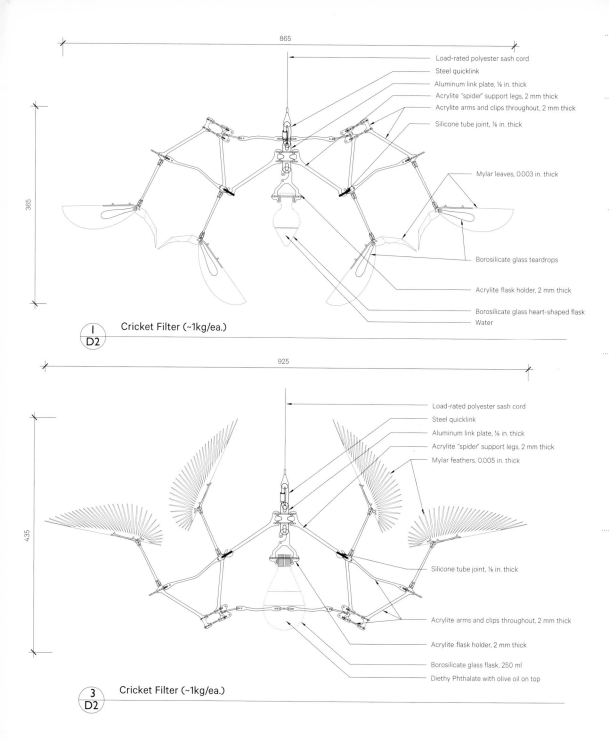

865

365

Load-rated polyester sash cord
Steel quicklink
Aluminum link plate, ⅛ in. thick
Acrylite "spider" support legs, 2 mm thick
Acrylite arms and clips throughout, 2 mm thick
Silicone tube joint, ⅛ in. thick

Mylar leaves, 0.003 in. thick

Borosilicate glass teardrops

Acrylite flask holder, 2 mm thick

Borosilicate glass heart-shaped flask
Water

1
D2 Cricket Filter (~1kg/ea.)

925

435

Load-rated polyester sash cord
Steel quicklink
Aluminum link plate, ⅛ in. thick
Acrylite "spider" support legs, 2 mm thick
Mylar feathers, 0.005 in. thick

Silicone tube joint, ⅛ in. thick

Acrylite arms and clips throughout, 2 mm thick

Acrylite flask holder, 2 mm thick

Borosilicate glass flask, 250 ml
Diethy Phthalate with olive oil on top

3
D2 Cricket Filter (~1kg/ea.)

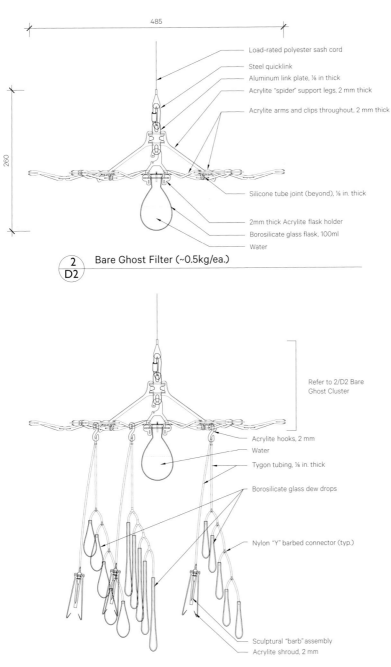

this and following spread: Architectural drawings of the complexity of the ecosystem

485

260

Load-rated polyester sash cord

Steel quicklink

Aluminum link plate, ⅛ in thick

Acrylite "spider" support legs, 2 mm thick

Acrylite arms and clips throughout, 2 mm thick

Silicone tube joint (beyond), ⅛ in. thick

2mm thick Acrylite flask holder

Borosilicate glass flask, 100ml

Water

2 / **D2** Bare Ghost Filter (~0.5kg/ea.)

Refer to 2/D2 Bare Ghost Cluster

Acrylite hooks, 2 mm

Water

Tygon tubing, ⅛ in. thick

Borosilicate glass dew drops

Nylon "Y" barbed connector (typ.)

Sculptural "barb" assembly

Acrylite shroud, 2 mm

4 / **D2** Bare Ghost Filter with Glass Dewdrops (~1kg/ea.)

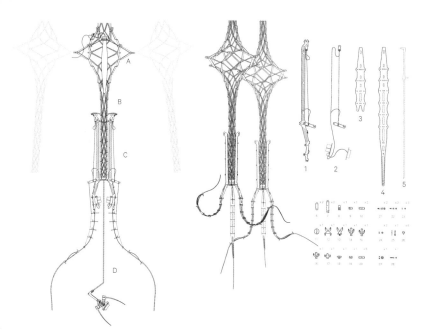

Tentacle Cluster

A Protocell flask with
 high-power LED reflex
B Thermoformed diagrid
 acrylic spar
C Tentacle with shape-memory
 alloy actuator
D Infrared proximity sensor
 with LED reflex

Tentacle Assembly

1 Aluminum sled mount
2 Shape-memory alloy
 actuated lever assembly
3 Copolyester tongue stiffener
4 Copolyester tongue
5 Isoprene polymer lash

pineal gland, the sternum, or the bundles of ganglia in elbows, knees, and palms. Those loosely tangled clusters of nerve centers suggest that one is a mongrel of approximately coordinated, roughly aligned species rather than one coordinated, fully conscious, integrated being. Efforts to engage the public are also inherent in the work via pursuit of a collective emplacement and renewed public architecture. If traditional public squares and meeting rooms carry weight from rigid kinds of nationalism and government, what kind of public gathering place might guide a space of contemporary gathering? The work implies renewed confidence about architecture functioning as a collective sanctuary. The scale is intended to work at an intimate, tribal level that gathers occupants throughout its porous, diffusive canopies and layered screen walls. By offering material turbulence as a primary design quality, *Epiphyte Chamber* moves from objective performance into realms of cultural iconography. Rather than strictly human-centered power, the ethics of mutual relations within wide and sometimes alien systems are implied by this work.

The design of *Epiphyte Chamber* emerged from the ongoing Hylozoic Series, an open and evolving series of experimental architectural installations that combine lightweight textilelike structures, responsive mechanisms, distributed computational controls, and chemical metabolic systems. It launched in 2007 with the presentation of *Hylozoic Soil* at Montreal's Musée des Beaux Arts, and features contributions from the many international collaborators in the Living Architecture Collective associated with the University of Waterloo. Using densely massed microprocessor-driven shape-memory alloy mechanisms and groves of glassware that hold colonies of synthetic biology, the works create immersive, interactive environments that move and breathe.

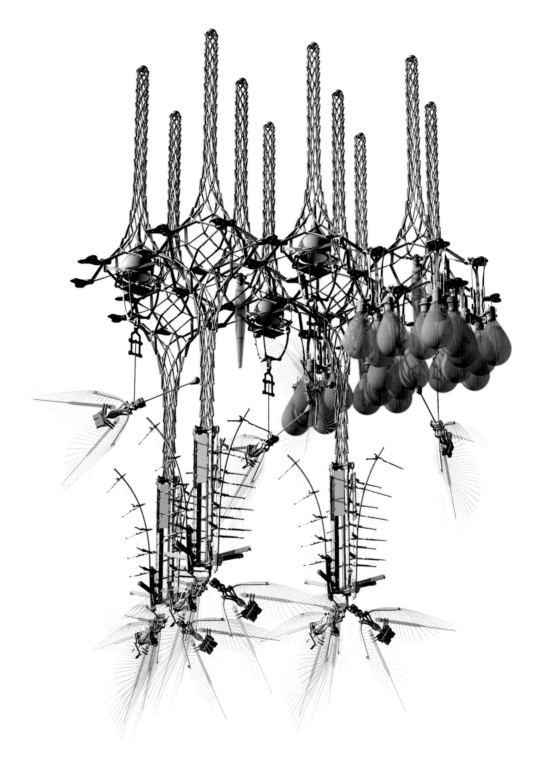

CONVENTIONS OF CONTROL
Michael Fox and Allyn Polancic

Interactive exhibit at the Hong Kong Polytechnic School of Design with students of interaction design. Gestural proximity controls the opacity of walls between users.

In the following project, Michael Fox and Allyn Polancic defined a catalog of architectural gestures for the remote control of dynamic architectural space, aiming to further characterize the notion of intuition as related to the creation of gestural languages. *Conventions of Control* used a Microsoft Kinect sensor and a series of typical, mechanically driven architectural fenestrations to discover the most referenced categories of relative intuition among test subjects prompted to invent gestures for specific architectural tasks. Their research points out the importance of body-space relationships with respect to our architectural environments. Gestural control is applicable to the following architectural systems: partition walls, skylights, doors, windows, lighting, and temperature. Responsive gestures, as an input, can easily evolve into biological sensory data and perhaps eliminate the need for a language in cases like temperature and light control. However, gestures might always be useful for architectural components like openings.

The project took an essential first step toward facilitating the field of architecture in playing a role in the development of an agenda for control. In creating a vocabulary for controlling dynamic architectural environments, Fox and Polancic's research builds on the

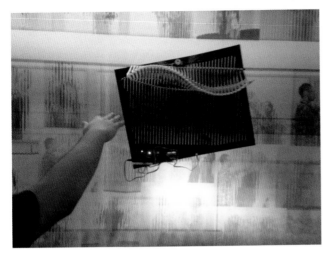

Student designs
utilize complex
gestural input
into mechanically
driven facade
prototypes

Architectural-scale
exhibit with gesturally
controlled navigational
screens that move
across a wall

state of the art of gestural manipulation, which exists in integrated touch- and gesture-based languages of mobile and media interfaces. The next step was to outline architecturally specific dynamic situational activities in order to explicitly understand the potential to utilize gestural control in systems that make up architectural space. A proposed vocabulary was then built upon the cross-referenced precedent of existing intuitive gestural languages as applied to architectural situations. The proposed gestural vocabulary was then tested against user-generated gestures in the following areas: frequency of invention, learnability, memorability, performability, efficiency, and opportunity for error. The testing was carried out through a test-cell environment with numerous kinetic architectural elements and the Kinect sensor to track gestures of the test subjects. In the context of the test cell, subjects were asked to control the various architectural components via gesture alone. Movement and body position were recorded and categorized, creating a taxonomy of user-generated intuitive gestures. Analysis revealed that intuitive gestures come from several different gestural languages that

we already know. The project aimed to discover the most referenced categories of relative intuition among test subjects prompted to invent gestures for specific architectural tasks.

If a user does not have a certain level of technical knowledge, he or she might not be able to understand how a component physically works, thus will not be able to operate it with even the simplest gesture. Furthermore, gestural control is meant to enhance an interactive environment, not burden its users. The ultimate purpose of a gesture is to send a signal. Whether this signal is used to open, close, or rotate, it is up to the programming of the computer and the mechanics of the architecture—not the user—to complete the given task.

As a means to explicitly understand gestural control relative to the physical architectural environment, a simple, fully functioning prototype was developed, consisting of a three-sided room with no ceiling. Placed within the wall were three motorized windows and a door, each of which could be independently controlled. The wall also had a single window that could rotate open or closed along the x-axis, and, additionally, a single partition wall that could move

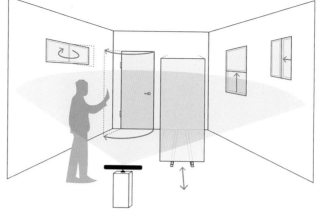

forward and backward on the y-axis. In terms of basic gesture recognition, the Kinect sensor was used to track the gestures of the test subjects. The sensor was an appropriate means of control that served to mimic what you would do in the real world as opposed to pressing buttons. In addressing the performance parameters of the prototype, the concept focused on several key strategies: (1) gestures, (2) physical movement, and (3) scale. The final objective of the approach was to create an innovative design that was minimally functional with the capability for evolving additional multifunctionality. Future applications may include motorized roller shades and/or a table that can slide out from the wall.

Exciting possibilities exist for augmented reality to include gestural languages that evolve into a type of dance, which in turn affects or transforms the surrounding environment of physically moving building components without physical labor while actually improving the body-space relationship that voice control and similar techniques continue to threaten. This relationship is extremely important in architecture, as people have a natural tendency to manipulate their surroundings through touch. The experience of architecture may be at first, and primarily, visual, but it is secondly and most crucially physical. We use our eyes to experience an architectural composition in 2-D, almost as a painting. However, the most engaging experience of architecture comes as the eyes negotiate distances, proportions, and materials in relation to the body. This is a physical relationship and is always subsequently verified by movement through and physical interaction with the building components. Physically interacting with the physical world is crucial to the way humans learn to understand reality, and the next phase of the project's gestural languages, utilized as control interfaces, should reflect this relationship and ideology.

Beyond this initial exploratory project aimed at understanding gestural interactions at an architectural scale, several subsequent projects have been developed by students using the more specific gestural capabilities of the Leap Motion sensor. The Leap Motion sensor allows for a very localized understanding of hand gestures in a much smaller sense space, approximately a two-foot virtual cube, above the

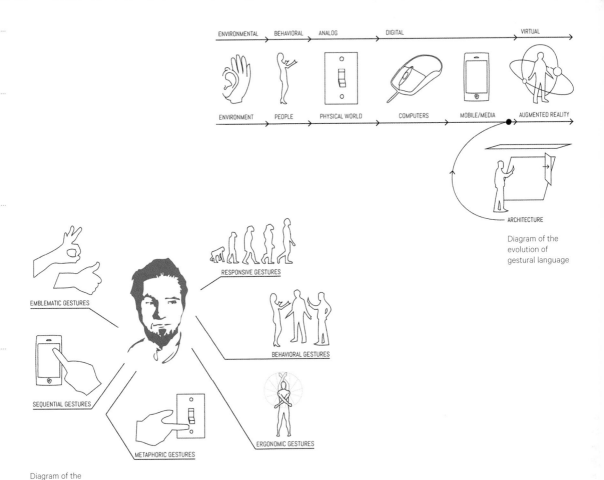

ENVIRONMENTAL → BEHAVIORAL → ANALOG → DIGITAL → VIRTUAL →

ENVIRONMENT → PEOPLE → PHYSICAL WORLD → COMPUTERS → MOBILE/MEDIA → AUGMENTED REALITY →

ARCHITECTURE

Diagram of the
evolution of
gestural language

EMBLEMATIC GESTURES

RESPONSIVE GESTURES

BEHAVIORAL GESTURES

SEQUENTIAL GESTURES

METAPHORIC GESTURES

ERGONOMIC GESTURES

Diagram of the
origins of intuitive
gestural languages

sensor. Numerous projects have been carried out at this level of specificity in gestural control, which bridges the general control and very specific localized control.

We conclude that the manipulation of physical building components and physical space itself is more suited to gestural manipulation by users instead of manipulation via device, speech, cognition, or other modes of operation. In the future it will be possible, if not commonplace, to embed architecture in interfaces to allow users to interact with their environments, and we believe that gestural language is the most powerful means of control through real physical interactions.

ALLOPLASTIC ARCHITECTURE
Behnaz Farahi Bouzanjani

According to architect Greg Lynn, "architecture is by definition the study and representation of statics. Architecture of the city must however embrace motion," because, as he claims, "classical models of pure static essentially timeless form and structure are no longer adequate."[6] How might we imagine a building that can amass an understanding of its users by studying their bodily gestures, expressions, behavior, and respond accordingly? How might we envision a space that possesses an interactivity based not simply on preprogrammed operations but on real-time feedback from its users? In other words, how might we envision a genuinely interactive space, the form and physical configuration of which can respond to and learn from its users? And how might such a space influence the ways we inhabit our environment, thereby changing the way we live? This case study investigates three interactive installation projects—*Alloplastic Architecture* (USC), *The Living, Breathing Wall* (USC), and *Breathing Wall* (downtown L.A.)—that address these questions. Within these projects lies the attempt to use emerging technologies, such as Microsoft Kinect and the Leap Motion sensor, and thereby understand new interaction scenarios and techniques that might inspire future research in this area.

Defining tensegrity parameters by coupling different materials with shape-memory alloy

The impulse behind all three investigations into gestural control was a desire to engage with the psychological benefits of an environment that can respond to—and therefore empathize with—human emotions through its capacity to adapt physically to the user. As such, the environment can be seen to overcome conditions of shock or alienation by accommodating the user. To design a biologically adaptive system, the observation of how living creatures constantly adapt to different external and internal stimuli in nature can offer considerable inspiration in terms of both their structural configuration and their process of adaptation. The issue is not simply how to create a system capable of changing, but how to research the quality of change and define the stimulus for adaptation. In times when the very concepts of nature and architecture are questioned not only in a philosophical dimension, but in the core of their biological materiality and relationship to humans, we need to reconsider how we are connecting to the environment. As the architect and theorist Neil Leach puts it: "Indeed the history of architecture can be read as a history of the relationship of human beings to their buildings. Attempts to relate the proportions of buildings to those of the human figure—from Vitruvius and before, through to Le Corbusier and other, more recent architects—are part and parcel of this history."[7] The projects here continue that strain not just in how we relate to our architectural surroundings, but in how that relationship is important for controlling them.

With *Alloplastic Architecture*, *The Living, Breathing Wall*, and *Breathing Wall*, the intent was to create spaces that could physically reconfigure themselves based, like any soft adaptive system in nature, on user movements. As theorist Sanford Kwinter observes, soft systems evolve by internal regulating mechanisms yet are always in collaboration with forces and efforts arriving from an outside source.[8] Therefore, the central focus in all the projects described here is the relationship between material behavior, form, and interactive systems of control.

In order to most effectively connect the human body to the environment, one might find a logic of behavior shared by the user and the surroundings. The result of this logic might be found through the use of dynamic tensegrity structures, designed to adapt to the behavior of the user. The advantage of this approach is

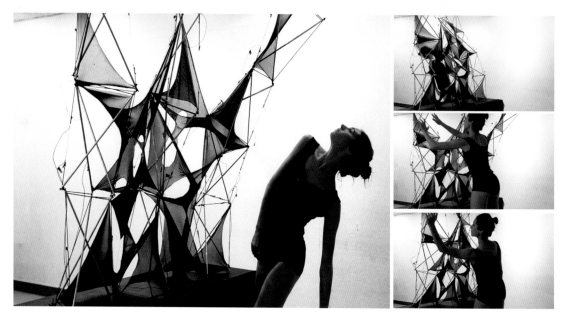

In *Alloplastic Architecture*, a performance artist dances with the structure, which reacts to her presence without any actual physical contact

that the human body itself is a form of tensegrity structure, with muscles and flexible tissues operating in a tensile capacity and bones and other rigid members operating mainly in a compressive capacity. If both the body of the user and the structure of the environment are governed by a similar logic of behavior, then the modeling of the behavior of one on that of the other will be all the more easy to engender.

Tensegrity structures are, of course, nothing new. The term *tensegrity* was coined by Buckminster Fuller, and successive generations of artists, architects, and engineers have developed the principle with an ever more sophisticated understanding of the behaviors of such structures. This project is one of the first, however, to investigate the potential for these structures to adapt and change their form. In *Alloplastic Architecture*, Bouzanjani used shape-memory alloy springs and other devices to operate as muscles capable of realigning a structure within a constant overall equilibrium. Other springs and expandable elements maintain the equilibrium by adjusting their length to compensate for an initial movement, thereby reconfiguring the entire structure.

Another device that was explored in this research project was the use of the Kinect motion sensor, which not only recognizes bodily movements and judges distance and depth, but also has the capacity to learn from users and adapt to them over time. Technically speaking, Kinect captures the bodily movement and Cartesian coordinates of the performer in the space. Based on digital information captured and processed with Processing software, various nodes of the structure are actuated with the help of an Arduino microcontroller, so the structure then bends toward or away from the performer. This project therefore addresses the potential of a reciprocal transformation between user and architectural element, whereby the environment influences the user and the user equally influences the environment.

Moving beyond capturing bodily movement of the users, *The Living, Breathing Wall* explores how a physical environment can change its shape in response to the speech recognition of users. The installation consisted of an eight-foot-by-twelve-foot (2.4 by 3.7 m) wood structure covered with tensile membrane fabric (spandex), aluminum strands, and a grid of fifty-six

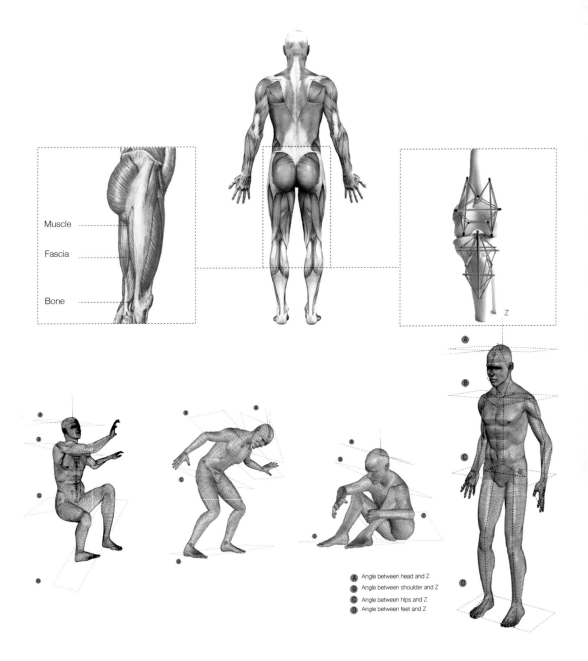

Muscle

Fascia

Bone

Z

A Angle between head and Z
B Angle between shoulder and Z
C Angle between hips and Z
D Angle between feet and Z

Kinect sensor captures the bodily movement and Cartesian coordinates of the performer in the space

Interactions
with *The Living,
Breathing Wall*

shape-memory alloy springs serving as actuators augmented with a brain (an Arduino MEGA 2560 and Kinect). The wall consisted of eight columns and six rows of eyelid-like aluminum strands that could open and close with the help of muscle springs. As each eyelid closed, it pushed the fabric surface outward to create various textures in the surface. The user interface (UI) in this project was a speech-recognition system (software) operating with Kinect (hardware), which recognized sentences and words as inputs and generated dynamic patterns as outputs. Indeed, if we are trying to imagine a future in which we have managed to liberate ourselves from

screens, voice-recognition systems seem inevitable. Speech recognition facilitates one of the most natural forms of human communication and can free up the hands and eyes for other tasks. Voice-based UI is potentially very convenient, as it obviates the need for physical gestures and would seem to be especially convenient for physically impaired users. Imagine, then, if we were to lie down on our couch and talk to our room so that it comes alive and offers us companionship.

In this case, the installation was programmed to respond to various key words in order to express specific feelings of happiness, sadness, excitement, and so on. A Kinect device had an

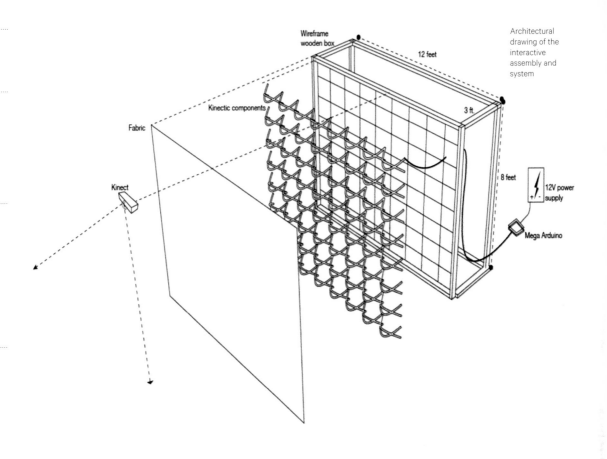

Architectural drawing of the interactive assembly and system

Wireframe wooden box

12 feet

3 ft.

8 feet

12V power supply

Mega Arduino

Kinectic components

Fabric

Kinect

array of four microphones built in and could therefore be used to recognize the words. It then relayed
instructions via an Arduino control board to the
shape-memory alloy–activated aluminum eyelid-like
components that moved in relation to a fabric
surface, creating various bumps or indentations
on the surface. Since each bump on the surface of
the wall acted as a pixel on an addressable grid,
the wall could potentially be controlled to display
ambient information on its surface, providing unob-
trusive content that is relevant to users. Then, if the
resolution of the pixels were increased sufficiently,
the wall could be used to transfer more detailed
content and information.

Breathing Wall showed how the use of remote sensing devices facilitates an empowered sense of embodiment in participants and offers an alternative way to remotely extend the range of our bodies within our physical environments

Mobile devices already use techniques grounded in touch- and gesture-based languages—swiping, clicking, dragging—as an intuitive mechanism of control. But can these techniques be used to control entire environments? *Breathing Wall* had two main objectives with this question in mind: First, it explored the potential for a gesture-based interaction with dynamic architectural space through the use of a Leap Motion sensor. Second, and more importantly, it explored the relationship between form and interactive systems of control in order to generate an empathetic relationship between users and their environment.

Breathing Wall consisted of an eight-by-twelve-inch (20.3 by 30.5 cm) wooden structure covered with a tensile membrane, direct current (DC) motors, and flexible PVC pipes. Each pipe was connected to two motors on each anchored point, which were capable of twisting and deforming the rod. Therefore, various surface formations could be generated with various commands from users. Early forms of remote devices might include remote controls for televisions or air-conditioning systems. However, the potential of telekinesis has expanded greatly with the advent of commercially available remote sensing devices, such as Kinect, that allow for both

Swipe Right

Point at wall

Swipe Left

Draw a circle with your finger

gesture-controlled and voice-controlled operating systems. Another product—using technology that is not dissimilar—is the Leap Motion sensor, which is more limited in its range as a motion sensor but more refined in what it can sense. Leap Motion can detect precise hand and finger motions at close range. It has a wide, 150-degree field of view and a z-axis for depth. The sensor detects the movements of the users as they move their hands in three dimensions, with various gestures such as swiping, drawing circles, and clicking on points in the space, in turn activating the surface of the wall. In the future, it may even be possible to design a direct interface that allows users to interact with their environment without any intermediary mechanism. Such interfaces will make control of our physical environment much easier and more intimate.

Together, these installations offer a vision of the future of our living spaces by demonstrating three interactive prototypes. They also engage in a series of interdisciplinary challenges, ranging from attaining a technological grasp of how such environments might change morphologically to gaining a psychological and neurological grasp of how human beings might themselves respond

to those changes and become active agents in remodeling and redesigning those environments. As the twenty-first century progresses, we are witnessing an increasing rise in the number of new technologies in our everyday lives. Some may be alarmed by this and find it an aspect of the dark side of our contemporary existence. Bouzanjani's projects seek to demonstrate and illustrate how interactive architecture has the ability to create nurturing conditions in which human beings engage in an empathetic relationship with their environment. Through such relationships, they might overcome their sense of alienation such that they become part of their environment and their environment, in turn, becomes part of them.

ACKNOWLEDGMENTS

This book was in the making for many years, while I was, in a sense, waiting on the profession. Clearly, such a book would not have been possible without the great work of many fantastic architects and designers who made the generous contributions of their amazing work. Thanks to those at Princeton Architectural Press, especially Jennifer Lippert for her support and Barbara Darko for both her editing skills and her constant nudging. I also greatly appreciate the many people who looked at various parts of the text, including Kenny On, Adriana Fuentes, and Jicheng Shen. I would like to thank my amazing wife, Juintow Lin, and little Juneau, Ori, and Io, who just like to see their names in books. Thanks to all those in my family, who mean so much. Thanks also to Bill Porter, who probably has no idea how much he inspired me at MIT. Finally, thanks for the many interesting conversations that have taken place, further shaping the book and inspiring its completion. I hope this book inspires new ways of exploring.

NOTES

FOREWORD

1 For more about the activities of the Interactive Architecture Lab at the Bartlett School of Architecture, University College London, see www.interactivearchitecture.org.

2 Nancy Lohman Staub, *Breaking Boundaries: American Puppetry in the 1980's* (Atlanta, GA: Center for Puppetry Arts, 1992), 20.

3 Rodney Brooks, "Natural Born Robots: Body Builders," *Scientific American Frontiers*, season 10, episode 2, directed by Graham Chedd and Andrew Liebman (1999; Alexandria, VA: PBS).

4 Harold B. Segel, *Pinocchio's Progeny: Puppets, Marionettes, Automatons, and Robots in Modernist and Avant-Garde Drama*, PAJ Books (Baltimore: Johns Hopkins University Press, 1995), 2.

INTRODUCTION

1 Neil Leach, ed., *Urban Architecture* 97 (September 2012): 8.

2 Behnaz Farahi Bouzanjani, "Alloplastic Architecture: The Design of an Interactive Tensegrity Structure" (presentation, ACADIA conference, University of Waterloo, Cambridge, Ontario, October 24–27, 2013).

3 Gordon Pask, "The Architectural Relevance of Cybernetics," *Architectural Design* (September 1969): 494–96.

4 John Frazer, *An Evolutionary Architecture*, Themes VII (London: Architectural Association Publications, 1995).

5 Adam Greenfield, *Everyware: The Dawning Age of Ubiquitous Computing* (Berkeley, CA: New Riders, 2006).

6 National Cable & Telecommunications Association, "Broadband by the Numbers—An Internet Built for All: Fast, Affordable and Competitive," *NCTA*, accessed March 29, 2015.

7 Stan Schneider, "Understanding the Protocols behind the Internet of Things," *Electronic Design*, October 9, 2013, http://electronicdesign.com/embedded/understanding-protocols-behind-internet-things.

8 Ibid.

9 Mat Honan, "Our Dream of the Connected Home Could Become a Nightmare," *WIRED*, August 6, 2014, http://www.wired.com/2014/08/connected-home/.

10 Jennifer Stein, Scott S. Fisher, and Greg Otto, "Interactive Architecture:

Connecting and Animating the Built Environment with the Internet of Things" (workshop, Internet of Things Conference 2010, Tokyo, Japan, November 29–December 1, 2010).

11 Ibid.

12 Hiroshi Ishii and Brygg Ullmer, "Tangible Bits: Towards Seamless Interfaces between People, Bits and Atoms," in *Proceedings of the ACM CHI 97 Human Factors in Computing Systems Conference* (1997): 234–41.

13 Eva Hornecker, "Tangible Interaction," *Interaction Design Foundation*, accessed August 16, 2013, http://www.interaction-design.org/encyclopedia/tangible_interaction.html.

14 Ian Daly, "Data Cycle: Behind MIT's SENSEable Cities Lab," *WIRED*, April 2011, http://www.wired.co.uk/magazine/archive/2011/04/features/data-cycle/page/3.

15 Bouzanjani, "Alloplastic Architecture."

EXHILARATE

1 William Zuk and Roger H. Clark, *Kinetic Architecture* (New York: Van Nostrand Reinhold, 1970), 9.

2 Kostas Terzidis, *Expressive Form: A Conceptual Approach to Computational Design* (London: Spon Press, 2003), 33.

3 Ibid., 45.

4 Cheng-An Pan and Taysheng Jeng, "Exploring Sensing-Based Kinetic Design for Responsive Architecture," in *Beyond Computer-Aided Design: Proceedings of the 13th International Conference on Computer-Aided Architectural Design Research in Asia (CAADRIA)* (Chiang Mai, Thailand: Association for Computer-Aided Architectural Design Research in Asia, 2008), 285–92.

COMMUNICATE

1 Nikos A. Salingaros, "Towards a Biological Understanding of Architecture and Urbanism: Lessons From Steven Pinker," *Katarxis*, September 2004, http://www.katarxis3.com/Salingaros-Biological_Understanding.htm.

2 Steven Pinker, "A Biological Understanding of Human Nature: A Talk with Steven Pinker," *Edge.org*, September 8, 2002, https://edge.org/conversation/a-biological-understanding-of-human-nature. Printed in John Brockman, ed., *The New Humanists* (New York: Barnes & Noble Books, 2003), 33–51.

3 Philip E. Ross, "Harman Cancels Out Road Noise, Without Headphones," *IEEE Spectrum*, last modified October 30, 2014, http://spectrum.ieee.org/cars-that-think/transportation/systems/harman-cancels-road-noise.

4 Ravi Mehta, Rui (Juliet) Zhu, and Amar Cheema, "Is Noise Always Bad? Exploring the Effects of Ambient Noise on Creative Cognition," *Journal of Consumer Research* 39, no. 4 (December 2012): 784–99, doi: 10.1086/665048.

5 Processing is available for download at https://www.processing.org/.

EVOLVE

1 John Frazer, *An Evolutionary Architecture*, Themes VII (London: Architectural Association Publications, 1995), 11.

2 Frazer, *An Evolutionary Architecture*, 13.

3 Richard Veryard, "Emergent Architecture," *Richard Veryard on Architecture* (blog), March 17, 2011, http://rvsoapbox.blogspot.com/2011/03/emergent-architecture.html.

4 Yoseph Bar-Cohen, *Biomimetics: Biologically Inspired Technologies* (Boca Raton, FL: CRC Press, 2005), 3.

5 Yoseph Bar-Cohen and Cynthia L. Breazeal, eds., *Biologically Inspired Intelligent Robots*, (Bellingham, WA: SPIE Publications, 2003), xiii.

6 Arnim von Gleich et al., *Potentials and Trends in Biomimetics* (Berlin: Springer, 2010), 19–22.

7 Toni Kotnik and Michael Weinstock, "Material, Form and Force," *Architectural Design* 82, no. 2 (March/April 2012): 104–11.

8 Ibid.

9 Blaine Brownell, *Transmaterial: A Catalog of Materials That Redefine Our Physical Environment* (New York: Princeton Architectural Press, 2005), 10.

10 Michelle Addington and Daniel Schodek, *Smart Materials and Technologies in Architecture: For the Architecture and Design Professions* (Oxford: Architectural Press, 2005), 83–95.

CATALYZE

1 Norbert Wiener, *Cybernetics: Or Control and Communication in the Animal and the Machine* (Cambridge, MA: MIT Press, 1948).

2 Sanford Kwinter, "Soft Systems," in *Culture Lab* 1, ed. Brian Boigon (New York: Princeton Architectural Press, 1993), 207–28.

3 See Martin Heidegger, "The Question Concerning Technology," in *Basic Writings*, ed. David Farrell Krell (New York: Harper Collins, 1993), 311–41.

4 Behnaz Farahi Bouzanjani, "Alloplastic Architecture: The Design of an Interactive Tensegrity Structure," in *ACADIA 2013 Adaptive Architecture: Proceedings of the 33rd Annual Conference of the Association for Computer Aided Design in Architecture*, ed. Philip Beesley, Omar Khan, and Michael Stacey (Cambridge, ON: Riverside Architectural Press, 2013): 129–36.

5 John Frazer, *An Evolutionary Architecture*, Themes VII (London: Architectural Association Publications, 1995), 7.

6 Greg Lynn, "An Advanced Form of Movement," *Architectural Design* 67 (May/June 1997), 54–59.

7 Neil Leach, "Adaptation," in *Unconventional Computing: Design Methods for Adaptive Architecture*, ed. Rachel Armstrong and Simone Ferracina (Cambridge, Ontario: Riverside Architectural Press, 2013), 130–31.

8 Kwinter, "Soft Systems," 218.

PROJECT CREDITS

EXHILARATE

MAY/SEPTEMBER
Year: 2014
Location: Indianapolis, Indiana, USA
Client/Institution/Sponsor:
Eskenazi Hospital of Marion County
Designer: Rob Ley / Urbana
Design Team: Loren Frances Adams,
Maxwell Miller
Fabrication: Indianapolis Fabrications
Structural Engineering: Nous
Engineering, Fink Roberts & Petrie
Photography Credits: Alan Tansey

TECHNORAMA FACADE
Year: 2002
Location: Winterthur, Switzerland
Client/Institution/Sponsor: Technorama,
the Swiss Science Center
Designer: Ned Kahn
Architects: Durig and Rami
Fabrication: Ned Kahn Studios
Photography Credits: Ned Kahn

WINDSWEPT
Year: 2011
Location: San Francisco, California, USA
Client/Institution/Sponsor:
Commissioned by the San Francisco Arts
Commission for a permanent installation
at the Randall Museum
Designer: Charles Sowers Studios
Fabrication: Spacesonic Precision Sheet
Metal, QC Facades
Structural Engineering: Hom-Pisano
Engineering
Installation: Rocket Science
Photography Credits: Bruce Diamonte,
San Francisco Arts Commission

REEF
Year: 2009
Location: New York, New York, USA
Client/Institution/Sponsor:
Storefront for Art and Architecture;
Supported by the AIA Upjohn Research
Initiative, the Graham Foundation for
Advanced Studies in the Fine Arts, the
AIA Knowledge Grant, and Interior Design
Educators Council Special Projects Grant;
Additional support and assistance
provided by Dynalloy
Designers: Rob Ley / Rob Ley Studio,
Joshua G. Stein / Radical Craft
Interaction Concept and Development:
Active Matter

Design Team: Timothy Francis,
Jonathan Wimmel, Elana Pappoff
Fabrication: Pylon Technical (motion
control software and custom electronics);
Peter Welch, Daniela Morales, Travis
Schlink, Joshua Mun, Darius Woo, Lisa
Hollywood, Rafael Rocha, Yohannes
Baynes, Phillip Ramirez
Photography Credits: Alan Tansey

COMMUNICATE

LIGHTSWARM
Year: 2014
Location: San Francisco, California, USA
Client/Institution/Sponsor:
Commissioned by Yerba Buena Center
for the Arts
Designers: Jason Kelly Johnson, Nataly
Gattegno, Ripon DeLeon
Design Team: Fernando Amenedo, Jeff
Maeshiro, Ji Ahn, Katarina Richter, Nainoa
Cravalho
Fabrication: MACHINIC Digital
Fabrication & Consulting
Photography Credits: Peter Prato

PLINTHOS
Year: 2014
Location: Athens, Greece
Client/Institution/Sponsor:
Commissioned by the ID10 Interior Design
Show
Designers: Franky Antimisiaris and
Branko M. Berlic / MAB Architecture,
Stavros Didakis
Photography Credits: Christos Drazos

BALLS!
Year: 2014
Location: London, England, UK
Client/Institution/Sponsor: Arup
Design Team: Alma-nac and Ruairi Glynn,
Tim Hunt, Francesco Anselmo
Structural Engineering: Interactive
Architecture Lab, Bartlett Faculty of the
Built Environment, University College
London
Software Programming: Felix Faire,
Francesco Anselmo
Project Manager: Craig Irvine
Photography Credits: Annabel Staff,
Tucker Productions (film stills)

MEGAFACES
Year: 2014
Location: Sochi, Russia
Client/Institution/Sponsor: MegaFon
Designer: Asif Khan Ltd.
Interactive Engineering: iart
Agency and Project Management: Axis
Structural Engineering: AKT II
Local Architect: Progress
Photography Credits: Asif Khan Ltd.

MEDIATE

AL BAHAR TOWERS
Year: 2008
Location: Abu Dhabi, UAE
Client/Institution/Sponsor: Abu Dhabi
Investment Council
Designer: Abdulmajid Karanouh, Aedas
Architects, Head of Ramboll Innovation
Design
Design Team: Pablo Miranda, Diar Consult
Structural Engineering: Arup
Photography Credits:

KFW WESTARKADE TOWER
Year: 2010
Location: Frankfurt, Germany
Client/Institution/Sponsor: KfW
Bankengruppe
Designers: Matthias Sauerbruch,
Louisa Hutton, Juan Lucas Young /
Sauerbruch Hutton
Project Leader: Tom Geister
Design Team: Jürgen Bartenschlag,
Peter Rieder, Marc Broquetas
Maduell, Christine Neuhoff, Barbara
Sellwig, Cynthia Grieshofer, Axel
Linde, Andrea Frensch, Lina Lahiri
Site Supervisors: Anton Bähr,
Angelika Fehn-Krestas, Claudius
Gelleri, Tanja Kausch-Löchelt, Timm
Knief, Daniela McCarthy, Mathias
Mund, Christiane Schmidt, Marcus
von der Oelsnitz
Electrical Engineering: Reuter
Rührgartner GmbH
Structural Engineering: Werner
Sobek Frankfurt GmbH & Co. KG
Photography Credits: Jan Bitter, Nosh

ECO-29
Year: 2013
Location: Hadera, Israel
Client/Institution/Sponsor:
Yosi Yahalom and Alon Talom

Interaction Concept and Design:
FoxLin Architects, Brahma Architects
Project Architect: Eran Shemish / Brahma
Architects
Electronics: Soundnine, Darius Miller
Photography Credits: Eran Shemish

SMART HIGHWAY
Year: 2012–present
Location: Province Brabant, City of
Eindhoven, City of Oss, and BKKC,
Netherlands
Client/Institution/Sponsor:
Joint venture with Studio Roosegaarde
and Heijmans Infrastructure; support
provided by Province Brabant, City of
Eindhoven, City of Oss, and BKKC
Designer: Studio Roosegaarde
Engineer: Heijmans Infrastructure
Photography Credits: Daan Roosegaarde

EVOLVE

HYGROSCOPE + HYGROSKIN
Year: 2012; 2013
Location: Paris, France
Client/Institution/Sponsor:
Commissioned by the Centre Georges
Pompidou, Paris, for its permanent
collection (first shown in the exhibition
Multiversités Créatives in 2012);
Commissioned by the FRAC Centre
Orléans (first shown as part of ArchiLab's
Naturalizing Architecture exhibition in
2013)
Designers: Achim Menges, Oliver David
Krieg, Steffen Reichert
Photography Credits: ICD University of
Stuttgart

BLOOM
Year: 2011
Location: Los Angeles, California, USA
Client/Institution/Sponsor: Materials &
Application Gallery, AIA Upjohn Research
Initiative, Arnold W. Brunner Award,
Graham Foundation Grant, University of
Southern California (USC) Advancing
Scholarship in the Humanities and Social
Sciences Program, USC Undergraduate
Research Associates Program, Woodbury
Faculty Development Grant; in-kind
donations from Engineered Materials
Solutions
Architect: Doris Kim Sung / DOSU Studio
Architecture

Consultants: Ingalill Wahlroos-Ritter
(design), Matthew Melnyk (structural
engineering)
Design Team: Dylan Wood (project
coordinator), Kristi Butterworth, Ali
Chen, Renata Ganis, Derek Greene, Julia
Michalski, Sayo Morinaga, Evan Shieh
Photography Credits: Brandon Shigeta,
Derek Greene, Doris Sung, Gerard
Smulevich

SHAPESHIFT
Year: 2010
Location: Zurich, Switzerland
Client/Institution/Sponsor: Starkart
gallery; Chair for Computer-Aided
Architectural Design, Institute of
Technology in Architecture, ETH Zurich
Designer: Manuel Kretzer
Design Team: Edyta Augustynowicz, Sofia
Georgakopoulou, Dino Rossi, Stefanie Sixt
Image Credits: Edyta Augustynowicz
Photography Credits: Manuel Kretzer

CATALYZE

EPIPHYTE CHAMBER
Year: 2013
Location: Seoul, South Korea
Client/Institution/Sponsor: Museum of
Modern and Contemporary Art, Seoul
Designer: Philip Beesley
Design Team: Martin Correa, Jonathan
Gotfryd, Andrea Ling, PBAI Studio, Sue
Balint, Matthew Chan, Vikrant Dasoar,
Faisal Kubba, Salvador Mirand, Connor
O'Grady, Anne Paxton, Eva Pianezzola,
Sheida Shahi, May Wu, Mingyi Zhou
Fabrication: Parantap Bhatt, Jessica
Carroll, Gelene Celis, Rayana Hossain,
Nada Kawar, Yonghan Kim, Pedro Lima,
Parham Rahimi Kearon, Roy Taylor, Cheon
Hari, Hanjun JoBoram Kim, Yonghan Kim,
Taehyung (Richard) Kim, Hyeon Min Lee,
Gyoung Hun Park, Sudam Park, Nuri Shin
Photography Credits: Philip Beesley

CONVENTIONS OF CONTROL
Year: 2012–2013
Location: Los Angeles, California, USA,
and Tianjin, China
Client/Institution/Sponsor: California
State Polytechnic University, Pomona
(CPP) and Tianjin University; support
provided by the Program of Introducing
Talents of Discipline to Universities (Grant
number: B13011)

Designers: CPP students: Allyn Polancic,
Rigo Gonzales, Sho Ikuda, Sizheng Chen,
Craig Aguilar, Cyrus Azari, Eric Mercado,
Kuniko Nickel, Jeffery Vanvoorhis, Jeff
Kerns, Junsin Miramontes, Michelle
Labininay
Photography Credits: Michael Fox

ALLOPLASTIC ARCHITECTURE
Year: 2012–2014
Location: Los Angeles, California, USA
Client/Institution/Sponsor: Mobile
and Environmental Media Lab (MEML),
University of Southern California (USC)
Designer: Behnaz Farahi Bouzanjani
Performance Artist: Nicole Ives
Photography Credits: Behnaz Farahi
Bouzanjani

Published by
Princeton Architectural Press
A McEvoy Group company
37 East Seventh Street
New York, New York 10003

Visit our website at www.papress.com

Project Editor: Barbara Darko
Designer: Jan Haux

Special thanks to: Nicola Bednarek Brower, Janet Behning,
Erin Cain, Tom Cho, Benjamin English, Jenny Florence,
Jan Cigliano Hartman, Lia Hunt, Mia Johnson, Valerie Kamen,
Simone Kaplan-Senchak, Stephanie Leke, Diane Levinson,
Jennifer Lippert, Sarah McKay, Jaime Nelson, Rob Shaeffer,
Sara Stemen, Kaymar Thomas, Paul Wagner, Joseph Weston,
and Janet Wong of Princeton Architectural Press
—Kevin C. Lippert, publisher

Front cover: *Lightswarm*, Future Cities Lab, 2014
Photograph by Peter Prato

Library of Congress Cataloging-in-Publication Data
Interactive architecture : adaptive world / Michael Fox, editor.
— First edition.
pages cm. — (Architecture briefs)
ISBN 978-1-61689-406-1 (paperback)
1. Architecture and technology. 2. Intelligent buildings.
3. Architecture—Technological innovations. I. Fox, Michael,
1967 August 22- editor.
NA2543.T43I59 2015
720'.47—dc23
                                        2015032533